chambre syndicale de la recherche
et de la production du pétrole
et du gaz naturel

comité des techniciens

commission exploration
sous-commission laboratoires de géologie

EVAPORITE DEPOSITS

ILLUSTRATION AND INTERPRETATION OF SOME ENVIRONMENTAL SEQUENCES

TIONS TECHNIP • 27, RUE GINOUX • 75737 PARIS CEDEX 15

ISBN 2-7108-0385-2

Printed in France
by Imprimerie Louis-Jean, 05002 Gap

FOREWORD

Evaporites are not normally hydrocarbon reservoirs, nor are they considered as possible source rocks. This simple observation singles out this work from the previous publications of the Committee of Technicians from *Chambre Syndicale de la Recherche et de la Production du Pétrole et du Gaz Naturel.* From this conclusion (that evaporites are neither reservoirs nor source rocks) there are fewer evaporitic cores, in comparison to those containing more hydrocarbons, to be found in the core-stocks of various companies, and the time spent by geologists working on site, or in the laboratory (which consequently influences the experience gained on these facies) will obviously be less than that spent dealing with porous formations. This is the basic reason why this work, in contrast with all previous publications of the Exploration Commission, has been undertaken not solely by oil geologists, who, in fact, are in the minority in the following list of authors : J. BOUROULLEC *(SNEA (P))*[1], P. BOUSQUET *(CFP),* G. BUSSON *(MNHN),* R. CUSSEY *(SNEA(P)),* Y. GUILLEVIN *(MNHN),* G. MAGLIONE† *(ORSTOM),* A.F. MAURIN *(CFP),* J.P. PERTHUISOT *(ENS),* J.M. ROUCHY *(MNHN),* D.J. SHEARMAN *(IC),* G. TRUC *(UL).*

The above, therefore, constitutes the first feature which singles this publication out from the rest.

The second proves to be of even greater importance. Most of the previous commission publications were produced as practical tools for use in the industry, aiming to standardize either study approaches or the presentation and processing of facts, or finally, how to express results. In nearly all of the cases, the representatives from different companies tried to reconcile their differing viewpoints, with the aim of facilitating exchanges and promoting progress. In this piece of work, we have had to be less ostentatious. We have, in fact, decided, for the moment at least, not to attempt to write a detailed account, or to classify the evaporites, or to give names to the sedimentary salt structures encountered with salts of greatest frequency (gypsum or anhydrite for example). We are simply attempting to offer the reader a few articles summarizing our experiences, articles in which (at the beginning at least) the illustrations should be more important than the text. We will try to justify this simple sequence of somewhat varied articles by explaining that they are intended to demonstrate different cases, describe vertical and horizontal series, illustrate structures and define the paleogeographical situation and significance of these facies. In short, the aim of these examples is

---

(1) *SNEA(P) : Société Nationale Elf Aquitaine (Production)*
CFP : Compagnie Française des Pétroles.
MNHN : Muséum National d'Histoire Naturelle.
ORSTOM : Office de la Recherche Scientifique et Technique d'Outre-Mer.
ENS : Ecole Nationale Supérieure.
UL : Université de Lyon.
IFP : Institut Français du Pétrole.
IC : Imperial College.

not to find a key to describe or interpret the evaporites, but rather to show the variation and learning wealth of facies, instead of an over-schematic representation, which could be made too hastily. Our project may be accused of lacking ambition, but, as it stands, there is no question of duplication with any existing publication, be it in France or abroad.

Is it necessary to justify *a posteriori* the creation of an Evaporitic Group by the Board of Technicians ? Or to allude to the importance of pre-evaporitic beds and the frequency with which they have served as source rocks, their capacity to create porosity (dissolution breccia), or to obliterate it in a host-sediment (carbonate or sandstone) through cementation, substitution or replacement, etc. ? Must we reiterate the imperviousness of caprock provided by evaporitic masses ; or finally recall that a large block of evaporites in a given area leaves its everlasting mark, often characterized by an unusual tendency towards instability (a later facilitation of creation of deposit centres, high zones, pinching-out, facies culminations, passage-beds, etc.) ?

In short, it should be apparent that although evaporites are normally neither source rocks nor reservoirs, they are, nevertheless, of considerable interest in the consideration of hydrocarbons. It is to the credit of the promoters of the Evaporitic Group that they recognized this fact at such an early date, particularly C. TEMPERE, President of the Exploration Commission and H.J. OERTLI, President of the Sub-Commission of Geological Laboratories.

This publication does not seem to be so obviously collective a work as numerous previous ones published by the *Chambre Syndicale de la Recherche et de la Production du Pétrole et du Gaz Naturel.* It is perhaps more necessary, therefore, to emphasize that its concept and completion were truly a collective act, each contribution having first been subject to discussion, criticism and sometimes to real joint efforts. In relation to this, the members of the group who do not figure in the list of authors must be mentioned particularly J. BAUER *(SNEA(P))*, J.P. BERTRAND *(IFP)* and L. YAPAUDJIAN *(SNEA(P)).* Their time and efforts expended on this work are all the more creditworthy since they were completely unselfish.

As a geologist outside the petroleum industry, I shall take the liberty of expressing to the reader a personal opinion of the previous publications of the Comité des Techniciens de la *Chambre Syndicale de la Recherche et de la Production du Pétrole et du Gaz Naturel.* Their utility and efficiency have, in my opinion, usually surpassed the modest objectives fixed by the authors. In this country of ours where, sadly, there is a very marked separation between industrial geologists and their counterparts in research and education, and where, for example, subsurface studies are almost exclusively undertaken by industrial geologists, previous publications by the Comité des Techniciens have sometimes led to the filling in of lacunas left gaping through lack of competent academics in the necessary fields.

May the thanks they deserve be accorded to these petroleum technicians, and long may they continue.

Georges BUSSON

CONTENTS

LIST OF FIGURES

Chapter I. Continental recent evaporites, Chad Basin (Africa)

Chapter II. Paralic recent evaporites, sebkha el Melah, Zarzis, Tunisia

Chapter IV. Evaporitic Middle Devonian successions, Alberta (Canada)

Chapter V. Sebkha, Lower Lias, southwestern France

Chapter VI. Evaporitic Miocene sequences, Sicily and Spain

Chapter VII. Petrography, Oligocene evaporites, Bresse, Valence, France

Chapter VIII. Diagenetic anhydrite, Tertiary sebkha, Iraq

Chapter XI. Depositional evaporitic sequences, diagenesis, Ludian, Stampian, southeastern France

LIST OF PLATES

INTRODUCTION

The initial idea on which this publication is based was the presentation of illustrations of depositional sequences in the evaporite environment. All members of the "Evaporite" group within the Committee of Technicians from *Chambre Syndicale de la Recherche et de la Production du Pétrole et du Gaz Naturel* endeavoured to select various aspects of these deposits, such as sedimentary structures, facies, petrography and diagenesis of the studied formations, this being carried out most often within the paleogeographic and/or sequential framework of these deposits.

Depending on the interest and personality of the individual, the reports may develop the sequential aspect, the petrographic aspect with and without relation to diagenesis. and the chemical aspect, which leads for example to the determination of the saliferous succession in the case of recent or sub-recent deposits.

In other respects, this publication embraces various fields of study, including the regional, sedimentary and stratigraphic fields.

Consequently, the choice of theme was a delicate matter: the subject could be presented from various angles. The plan adopted, therefore, does in no way claim to solve these difficulties, but partially introduces some form of classification.

The first part basically concerns continental or margino-littoral deposits of recent to sub-recent age, and consists of three chapters. The last of these chapters deals with the relation between ancient and recent sedimentation by way of comparison between recent and mesozoic deposits.

The second part contains chapters IV to VII which treat different examples of sequential studies in macro- and microlithology, from the Devonian to the Miocene, on varied domains of sedimentation between the littoral marine and probably far greater depths.

Chapter VII of this second part presents a more specific diagenetic aspect (as do the last four chapters of the third and final part) while relocating these phenomena in the sequence of the deposit (chapters VIII to XI).

An alphabetic index records the names of fields and sedimentary structures, mineralogy and petrography, stratigraphy, formations and geographic localities studied, as well as the names of authors with selective references to chapters, pages, figures in text and tables.

As G. BUSSON emphasises in the foreword this publication surveys the different projects undertaken during 1975-1977 by a group of researchers brought together in the Committee of Technicians from *Chambre Syndicale de la Recherche et de la Production du Pétrole et du Gaz Naturel.* This committee is honoured by the very fruitful collaboration of Professor D.J. SHEARMAN of Imperial College, London, to whom it would like to express its most sincere thanks.

In other respects, let us hope that the next object of study on evaporites adopted by this group of researchers – the rounding-up of different methods of approach to the study of evaporite sediments – may supply new elements of greater didactic value.

first part

RECENT TO SUBRECENT SEDIMENTATION EXAMPLES

I

AN EXAMPLE OF RECENT CONTINENTAL EVAPORITIC SEDIMENTATION THE CHADIAN BASIN (AFRICA)

G. MAGLIONE *

Plates 1 to 9

The Chadian centroclinal is a broad endoreic basin covering nearly three million square kilometres. Its current important morphological features result, for the most part, from climatic alterations of the Quaternary. To the south, following the Precambrian mass of the Central African Republic, the great flood plains meet delta sewage and fluvio-lacustrine zones, laid down during "pluvial" periods. To the north of the 13° latitude, the permanent Kanem erg bears witness to an early period of arid climate (see Fig. 1).

In the direction NNW-SSE, the interdunary depressions of this erg trapped fine sediments of clays and diatomites during a period when Lake Chad was in expansion at the time of the Holocene (about 6 000 years B.P.) when it covered up to 350 000 square kilometres.

Present day Lake Chad forms a visible receptacle for the tributaries which drain the basin. It does not, however, occupy the lowest lying ground, which is to be found 500 km northeast of the lake at 170 m above sea level, known as the "Pays Bas" (Lowlands). The sinking of chains of dunes in the lake obstructs the flow of water towards this region through a fossil outlet, the "Bahr-el-ghazal".

The subdesert climate of this region is characterised by a long, dry season alternating with a short, wet season (300 mm between July and September), high temperatures (an annual average of 28°C) and a very dry atmosphere (average annual level of humidity : 36 %). The level of evapo-transpiration is very high, 2 200 mm/year.

The most important feature, therefore, is the very distinct **high level of evaporation**.

In a year of average precipitation, the Chari, the main tributary (swollen between Logone and N'Djamena), carries 40 000 million cubic metres of water to Lake Chad, transporting two million tonnes of salt in ionic form (mostly carbonates of alkaline earths) and one million tonnes of dissolved silica.

In an average year, Lake Chad covers an area of about 20 000 square kilometres, its altitude varies between 280 and 282 m, its average depth is about 3.4 m and its volume about 58 000 million cubic metres.

(*) Formerly at Laboratoire de Géochimie, Office de la Recherche Scientifique et Technique d'Outre-Mer, Boîte Postale 1386, Dakar, Sénégal.

Hydrogeological studies have shown that the lake's expanse, with its altitude of 282 m, occupies a high point in the hydrological system. It is worth noting that on its northern edge, the lake is skirted at a distance of several kilometres from its banks by a drainage waterway which, at its lowest point, is at an altitude of 275 m, that is 5 to 7 m below the actual average level of the lake.

Lake Chad is, in fact, only a "transit camp" for the waters, and for most of the soluble salts brought down to it by its rivers. It can therefore be considered as a "see-page lake", as coined by anglo-saxon authors. Various estimates made from hydrons and saline check-ups have permitted losses through evaporation to be evaluated at a level of 92.5 % and those by infiltration at 7.5 % of the total annual liquids received (rivers and rain). Most of the losses through infiltration occur on the north east edge of the lake, precisely where the interdunary depressions at present form the base of an evaporitic sedimentation.

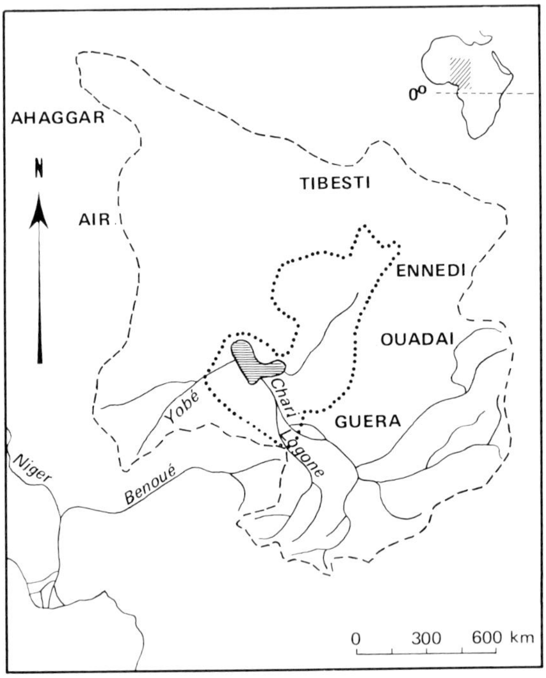

Fig. 1. – Location of the Chad Basin.
——— Basin boundary.
••••••• Lake shore line at 320 m (approximately 6000 yrs B.P.).

The waters of the Chari are slightly mineralised bicarbonated sodi-calcic (50 mg/l). Lake Chad acts as a pre-concentration basin, the waters becoming more concentrated through evaporation from south to north. On the northern shores they are 10 to 15 times more loaded than at their outlet in the Chari delta.

Apart from this quantitative evolution, the lake is also the base for qualitative change resulting from chemical and biochemical sedimentation. Taking the sodium ion (the most inert) as tracer, only 16, 40, 46, 56 and 64 % of the incoming annual total of silica, calcium, magnesium, bicarbonate and potassium respectively, are reported to have drained off Lake Chad.

Lake Chad, therefore, acts as a basin for **pre-concentration** as well as for **load-shedding** of various salts.

In the interdunes the waters become more concentrated in two ways :
(a) By direct evaporation in the interdunary lakes.
(b) By capillary evaporation, in the case of a submerged nappe.

The first case leads to a lateral saline zoning ; the salts spread in approximately concentric aureoles, the most soluble nearest the centre. This zoning results from salt crystallisation, beginning salt brines during the drying out of interdunary lakes. This will be refered to, then, as **syngenetic** paragenesis.

The second instance involves both lateral and vertical zoning within the sediment, by capillary pressures. This type of paragenesis will thus be **diagenetic**.

In the same interdune, both processes may have been in operation several times over, alternating according to fluctuations in the level of Lake Chad over the last twelve millenniums. The present day distribution of salts in profiles is but the **result of both of these processes.**

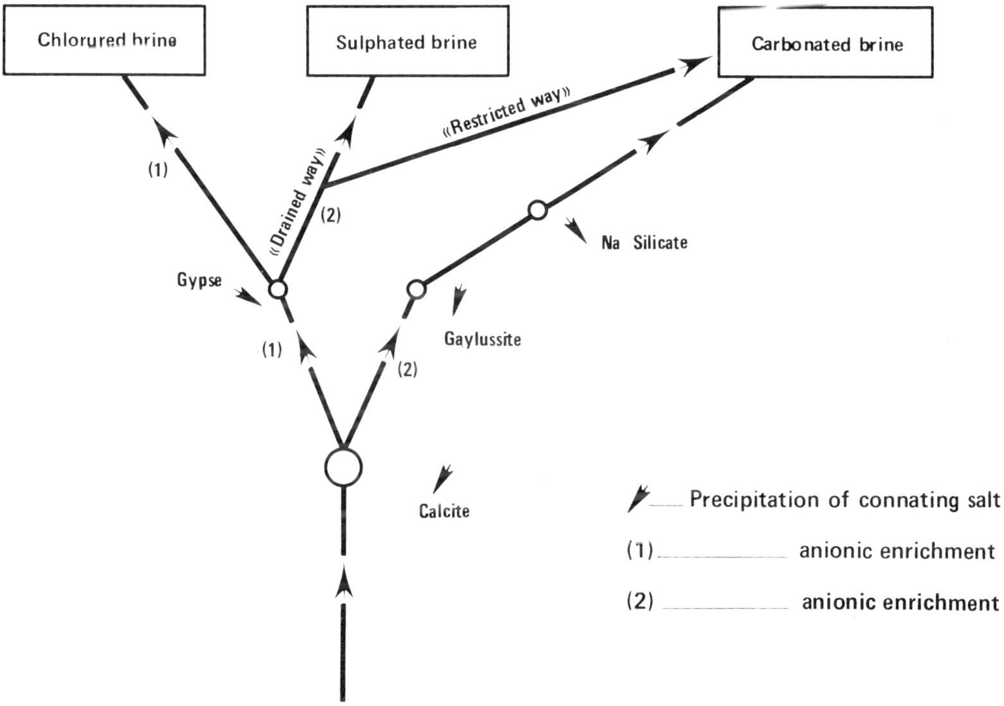

Fig. 2. – Dichotomic evolution of water influenced by evaporation and successive crystallisations (modified by L.H. Hardie and H.P. Eugster, 1970).

In both examples, the paragenesis is the same, made up, for the most part, of carbonated alkaline salts such as gaylussite, $Na_2Ca(CO_3)_2 . 5\ H_2O$, nahcolite, $NaHCO_3$, natron, $Na_2CO. 10\ H_2O$, thermonatrite, $Na_2CO_3 . H_2O$, trona, $Na_3H(CO_3)_2 . 2\ H_2O$ with which a little halite (NaCl) and gypsum are mixed. In the meantime the two mechanisms give birth to two generations of mineralogical characters identifiable in profiles (Fig. 2).

As well as true salts, attention should be drawn to the neoformation of silicates such as mordenite, magadiite ($NaSi_7O_{13}(OH)_3 . 3\ H_2O$), kenyaite ($NaSi_{11}O_{20} . 5(OH)_4 . 3\ H_2O$) and lastly kanemite ($NaH(Si_2O_4(OH)_2 . 2\ H_2O$), a new mineral found to this day only in the *natronières* of the Kanem.

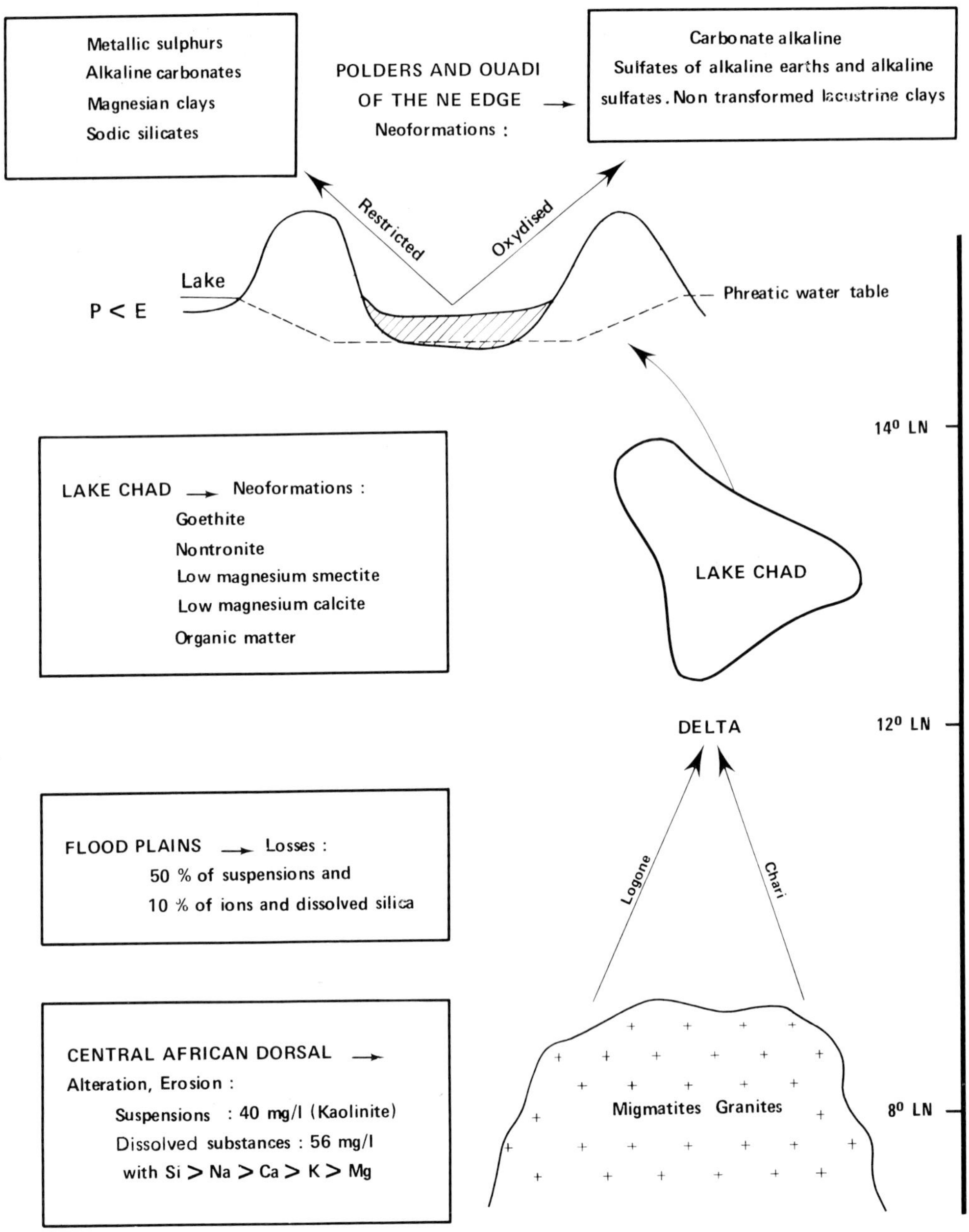

Fig. 3. – Surface dynamics of the Chad Basin.

Another original feature occurs in the drainage and aeration of interdunary depressions. In the poorly drained, poorly oxygenated environments which occur in greatest frequency, the sulphate ions are held back in the state of alkaline and metallic sulphurs by bacterial reducing mechanisms : this paragenesis is **carbonated sodic.**

In well drained, well aerated environments, sulphates may react mineralogically, combining chlorides, carbonates already having been caught upstream. This paragenesis is now, therefore, predominantly **sulphated sodic** with minerals such as gypsum, bloedite ($Na_2Mg(SO_4)_2.4H_2O$), mirabilite ($Na_2SO_4.10\ H_2O$) or in its dehydrated form, thenardite (Na_2SO_4) joined halite (NaCl) and with rarer minerals, such as northupite ($Na_2CO_3.MgCO_3.NaCl$).

In time, across the whole Chadian Basin, a reduction in the limestone content of the lacustrine sediments has been observable from the Plio-pleistocene age up to the present. This could result from the beginning of a new endoreic cycle corresponding to the restricted environment of the basin, the carbonate having been the first to precipate considerably.

To conclude, the present day surface dynamics of the material in the Chadian Basin can be synthetized by Figure 3.

SELECTED REFERENCES

Carmouze (J.P.) and *al.*, 1975, "Aspects sédimentologiques actuels d'un bassin continental endoréique : le bassin tchadien". *Actes du IXe Congr. Intern. Sédimentologie,* Nice, 6-13 juillet 1975.

Maglione (G.), 1974, *Géochimie des évaporites et silicates néoformés en milieu continental confiné. Les dépressions interdunaires du Tchad, Afrique.* Thèse, Univ. Paris VI, 331 p.

II

SEBKHA EL MELAH NEAR ZARZIS A RECENT PARALIC SALT BASIN (TUNISIA)

J.P. PERTHUISOT *

Plates 10 and 11

The Sebkha el Melah, situated on the eastern coast of South Tunisia (Fig. 4) is a flat, roughly rectangular plain with an area of about 150 km^2, its altitude varies between + 1 m and – 1 m, but most of the sebkha is actually below sea level. On its continental margins, the sebkha is cut into a series of plateaux which were shaped during successive climatic and eustatic changes in the Quaternary era. The basin is separated from the sea by the Tyrrhenian dunes, which are cut by a dry channel, with a sill at 1 m above sea-level. The substratum consists mainly of Mio-pliocene gypsum and clays. The basin's infilling can be regarded schematically as comprising two sequences :

1. The lower sequence is a fluvio-marine layer, overbedded by a fossil-bearing marine layer – both attributable to the medio-wurmian transgression (about 35 000 B.P.) which seems to have reached a level several metres below present day sea level.

2. The complete upper sequence is the most important part of infilling and is Flandrian. It contains, working upwards (Figs. 5 to 10) :

(a) An organo-detrital layer : carbonated, fossil-bearing and marine.
(b) A layer of hypermagnesian carbonates (magnesite and huntite), rich in gypsum lenses.
(c) A mainly gypseous sulphated layer (gypsum transformed locally into polyhalite), absent from deepest parts of the basin.
(d) A halite layer with a maximum thickness of 30 m, containing a little gypsum and/or polyhalite, particularly near the bottom.
(e) A thin layer of black salty mud, containing magnesite and syngenetic polyhalite, locally.

All these units "rest" inside each other, forming concentric aureoles.

(*) Laboratoire de Géologie, École Normale Supérieure, 46 rue d'Ulm, 75005 Paris, France.

SELECTED REFERENCES

Perthuisot (J.P.) and Florida (S.), 1973, *Carte géologique de la Sebkha el Melah et de ses bordures.* Trav. lab. Géol. Ecole Normale Sup., **8**, carte 1/50 000 & notice, 23 p.

Perthuisot (J.P.), 1975, *La Sebkha el Melah de Zarzis. Genèse et évolution d'un bassin salin paralique.* Trav. Lab. Géol. Ecole Normale Sup., **9**, 252 p.

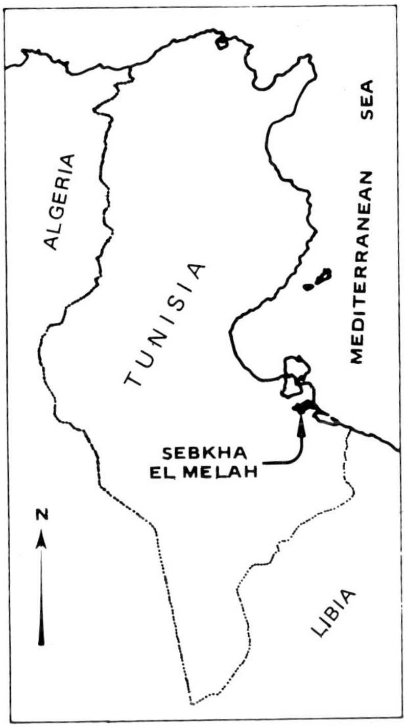

Fig. 4. – Situation of the Sebkha el Melah.

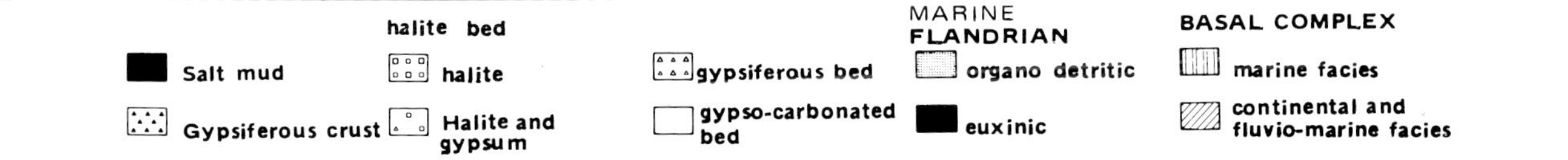

Fig. 5. – Schematic section of the sedimentary infilling of the Sebkha el Melah.

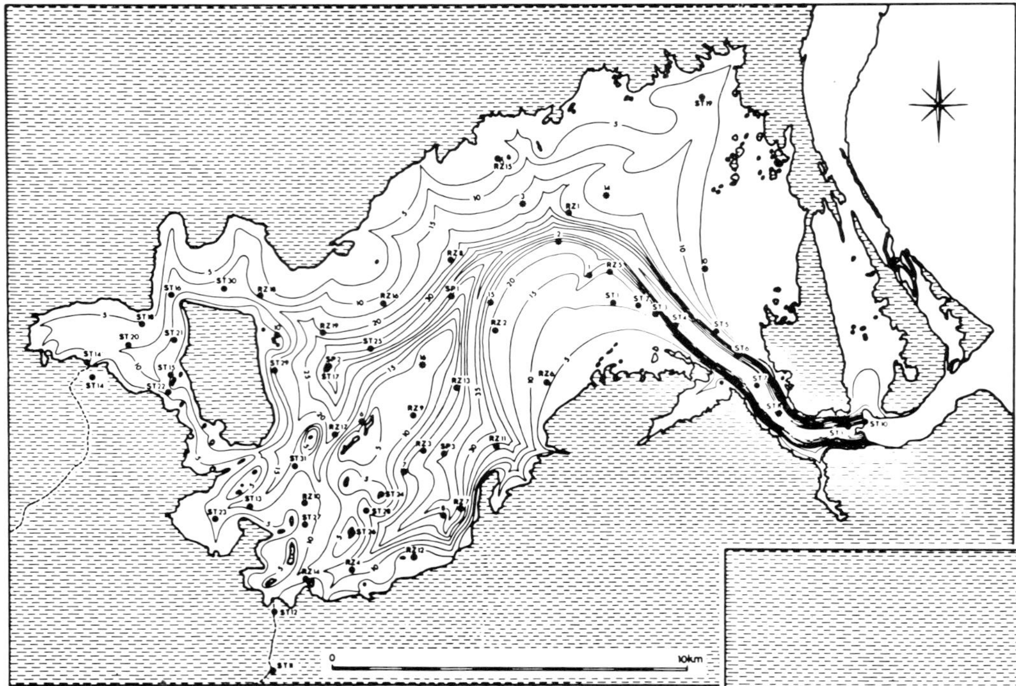

Fig. 6. – Isobathic map of sedimentary infilling of the Sebkha el Melah

In black the position of stromatolithic formations.
The basin was hollowed out during Wurm II pluvial stage by strong wadis issuing from the mountains of South Tunisia. The Medio-wurmian transgression (Wurm II-III interstade) floods the basin, forming first an estuary, then a lagoon. The transgression reaches its peak around 35 000 years B.P., and leaves a lacustrine deposit in the basin – fluvio-marine to marine in the channel zone – where a sill forms.
During Wurm III and IV the sea withdraws, separating the basin from the sea and turning it into a continental sebkha. An aeolian sill forms on its eastern margin.

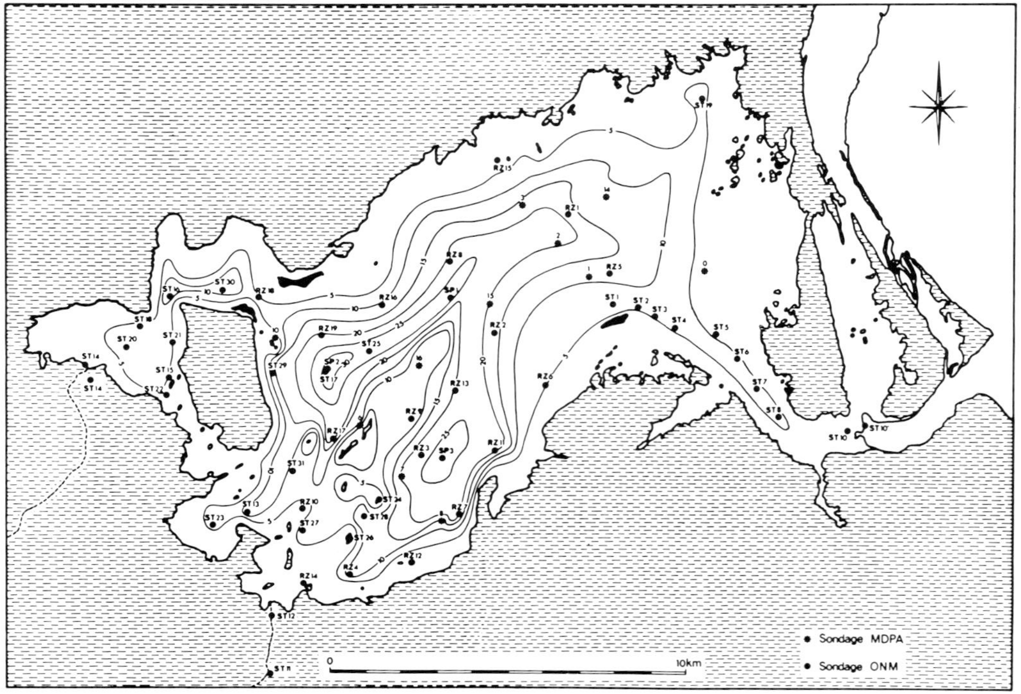

Fig. 7. – Isobathic map of the Flandrian sequence in Sebkha el Melah.

The Flandrian transgression floods the sill around 8 000 B.P. and the basin turns into a "Bahira" (small sea) where fossiliferous sandy and marly carbonates are deposited. The sill strengthens in the channel ; the sea level reaches its peak around 6 000 B.P. (at about + 2 m) and then lowers as the climate becomes drier. The communication with the sea becomes more and more restricted until the basin becomes totally confined, leading to evaporitic sedimentation.

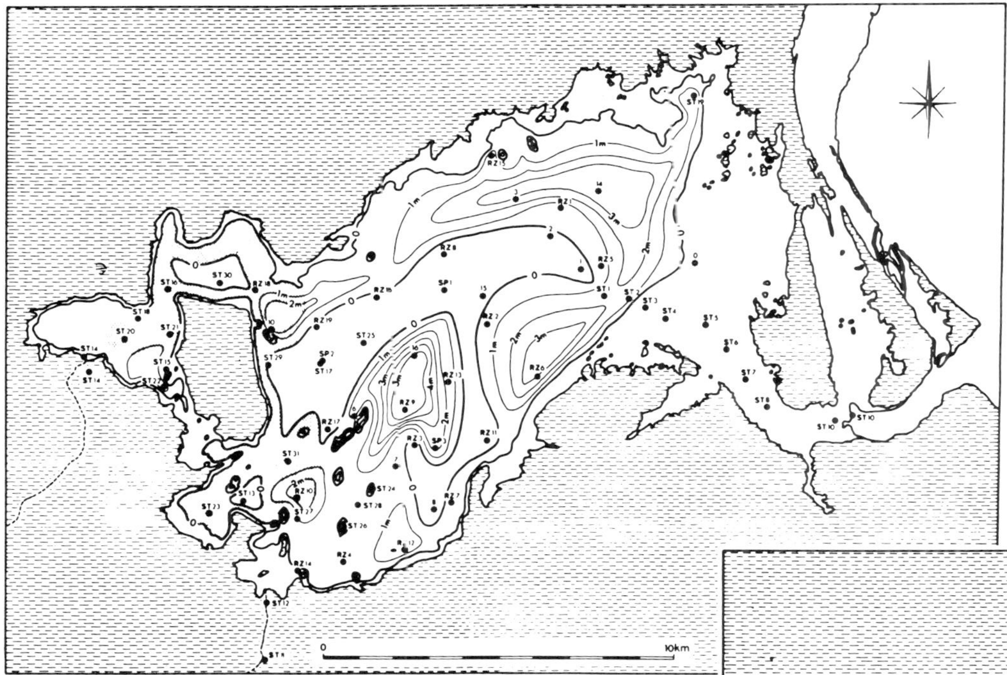

Fig. 8. – Isopach map of the sulphate layer (mainly gypsum). (Dashed area : polyhalitised gypsum).
The map clearly shows that gypsum is restricted to the shallow parts of the basin (margins, middle ridge). The absence of it in the deepest parts of the basin is caused by the activity of sulfo-reducing bacteria (very probably *Desulfovibrio*). These discard ions of CO_3^{--} and CO_2 into the environment and thus aid carbonate sedimentation. During halite precipitation in the basin, brines become enriched in Mg^{++} and K^+. After this periplus they sink to the bottom of the basin, react on gypsum and produce polyhalite $K_2CaMg(SO_4)_4\ 2\,H_2O$. This mechanism explains the distal location of this diagenetic polyhalite.

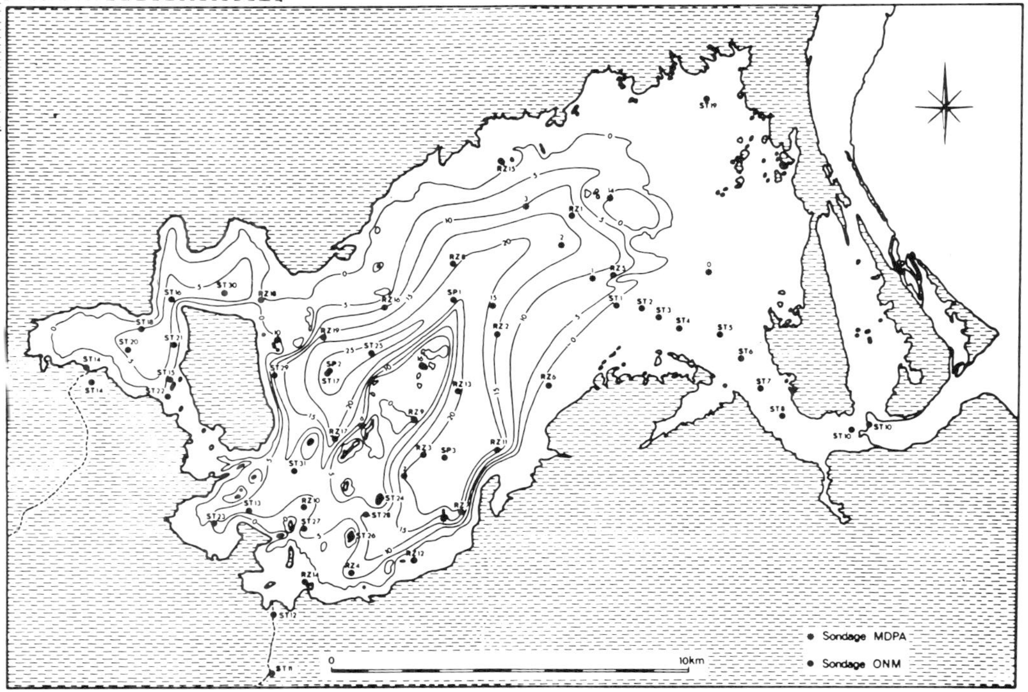

Fig. 9. – Isopach map of the polyhalite layer.
A strong rock salt lens forms in the deepest parts of the basin, but towards the continental edge, its position is somewhat eccentric.

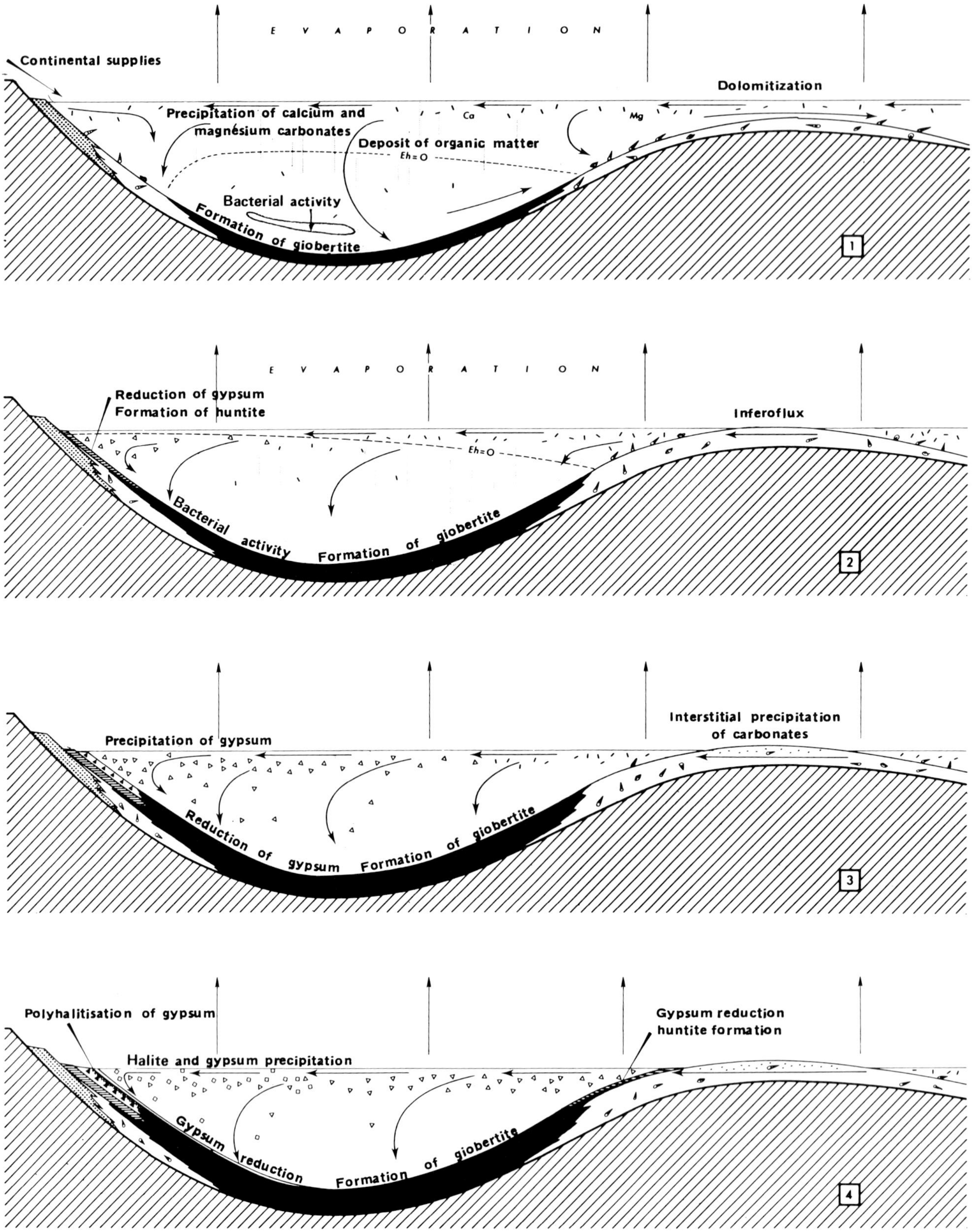
E V A P O R A T I O N
Continental supplies
Dolomitization
Precipitation of calcium and
magnesium carbonates
Ca
Mg
Deposit of organic matter
Eh=O
Bacterial activity
Formation of giobertite
1
E V A P O R A T I O N
Reduction of gypsum
Formation of huntite
Inferoflux
Eh=O
Bacterial activity Formation of giobertite
2
Precipitation of gypsum
Interstitial precipitation
of carbonates
Reduction of gypsum Formation of giobertite
3
Polyhalitisation of gypsum
Gypsum reduction
huntite formation
Halite and gypsum precipitation
Gypsum reduction Formation of giobertite
4

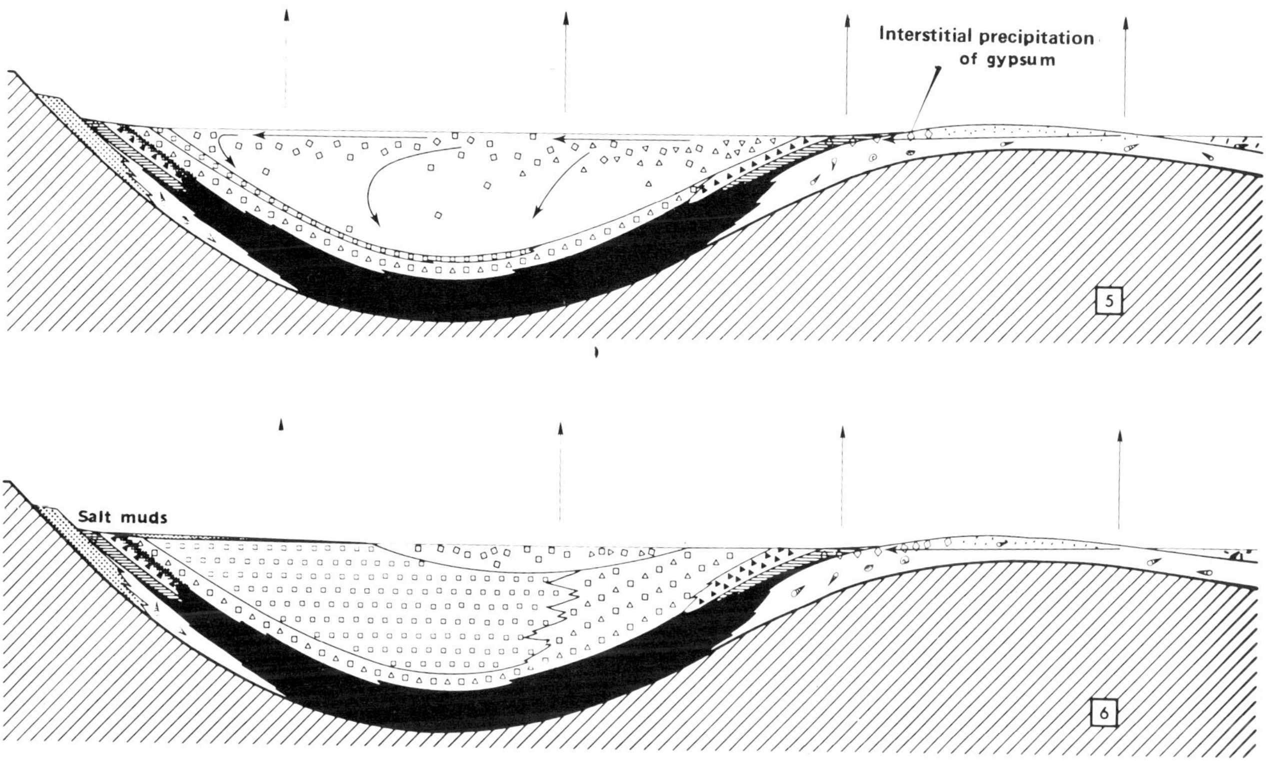

Fig. 10. – Development of Flandrian sedimentation in Sebkha el Melah.

1. Flandrian maximum. The lagoon is directly supplied with sea water. Deposition of the organo-detrital layer. From this stage onwards, a water lens with an increased Mg/Ca ratio forms at depth. In the deepest, euxinic parts of the basin, bacterial activity leads to magnesite formation ($MgCO_3$).

2. The sea level lowers, the basin starts to be supplied by underflow through the channel sediments. Confinement increases and a salinity gradient becomes established over the basin. Gypsum starts to precipitate in the western part of the basin. Sulfo-reducing bacteria act on it and its place is taken by magnesian carbonates (huntite, $CaMg_3(CO_3)_4$).

3. The sea level continues to fall, as does the water level in the basin. The amount of underflowing water towards the basin increases. The salinity gradient is strengthened and moves towards the sea. Gypsum precipitation gradually occurs over the whole basin. In the distal part, bacterial action cannot control gypsum precipitation, so gypsum persists. However, in the deepest parts of the basin, gypsum reduction by bacteria still prevails over precipitation.

4. All these processes increase. Halite starts to precipitate in the western zone (the distal or continental part), where previously deposited gypsum is transformed into polyhalite.

5. Halite precipitation takes hold over the whole basin, on its own in the west, in the east still mixed with gypsum.

6. The basin is completely filled with sediments and the evaporitic sequence is gradually being covered by a continental detrital mud sheet.

The final pattern of "rested", concentric evaporitic layers conceals an instantaneous spread with longitudinal segregation of salt precipitation zones according to the basin's salinity gradient.

III

SEBKHA FACIES EVAPORITES

D.J. SHEARMAN *

Plates 12 to 17

Coastal sebkhas can be thought of as being the desert zone analogues of the salt marshes of more temperate regions, but they differ in the absence of vegetation and that they can be the sites of interstitial deposition of evaporite minerals. The latter are precipitated from the marine-derived ground-water brines as they become increasingly concentrated as a result of capillarity and evaporation. The coastal sebkhas of the Sheikdom of Abu Dhabi, United Arab Emirates, are an interesting example, because they are the sites of present day emplacement of nodular anhydrite on a regional scale. The sebkhas are prograding supratidal flats which lie behind a complex of shallow water, recent marine carbonate sediments. In simplest terms, the sediments of the sebkhas comprise a vertical succession of lagoonal or subtidal carbonates which pass up into intertidal algal mats, which are in turn overlain by the sediments of the supratidal facies. The succession records a simple regressive cycle of carbonate shoreline sedimentation. Small crystals of gypsum grow abundantly in the intertidal algal mats and, a short distance inland of normal high tide level, series of anhydrite nodules grow by displacement in the sediments of the supratidal facies. Emplacement of the evaporites is accompanied by dolomitization of the fine grained aragonite present in the intertidal and supratidal sediments.

Comparable features are seen in ancient carbonate/anhydrite cycles, such as in the Lower Purbeck rocks at Warlingham, Surrey, England.

(*) Department of Geology, Imperial College of Science and Technology, Prince Consort Road, London SW7 2BP, Great Britain.

second part

EXAMPLES OF FORMER EVAPORITIC DEPOSITIONAL SEQUENCES

IV

CARBONATE, ANHYDRITIC AND SALIFEROUS SEQUENCES IN THE MIDDLE DEVONIAN OF WELLS IN NORTH-EASTERN ALBERTA (WEST CANADA)

G. BUSSON *

Plates 18 to 30

The so-called Upper Elk Point subgroup is a series of carbonate and evaporite layers, dated approximately as Middle Devonian, and covering part of the Western plains of Canada and the United States (Alberta, Saskatchewan, Montana, North-Dakota, etc.). The series is made up of :

1. At the base, limestones called "Keg River" formation in Alberta, and "Winnipegosis" in Saskatchewan.
2. Above, evaporites called "Muskeg" formation in Alberta and "Prairie-Evaporites" in Saskatchewan.
3. At the top, more argillaceous beds called "Watt Mountains" in Alberta and "Red Beds" or "Dawson Bay" in Saskatchewan.

The extent, variety and spread of the different facies encountered are outstanding : there are almost 1 500 km between the open marine facies which are equivalent to the evaporites (marine marls and "Pine Point" limestones in the North-West) and the bottom end of the basin in the South-East. Throughout these 1 500 km, beyond the marine marls and limestones already mentioned, we can observe series with a predomination of dolomite, then anhydrite, then rock salt, followed by series with outstanding interbedded potash layers (Saskatchewan) and finally detritics at the extreme edge of the basin.

The boundaries of Upper Elk Point are thus stratigraphically determined by the wall of Keg River Winnipegosis limestones at the base, and by the top of the Watt Mountains formation or Red Beds formation at the top. On the whole, most specialists have considered these two contact surfaces to be synchronous. The body they enclose, although imprecisely and even with some doubt dated as Middle Devonian,

(*) Laboratoire de Géologie, Muséum National d'Histoire Naturelle, 43 rue de Buffon, 75005 Paris, France.

is thus a typical parastratigraphic unit (Busson, 1972) : i.e. a sequence of beds approximately synchronous, but poorly placed in the universal stratigraphic scale.

The unusual development and spread of the facies are not the only interesting features which make the study of these layers attractive. The very slight tectonic distortion affecting these salt beds should also be noted, for, in these regions, the Devonian shows a dip of less than 1° ! This explains the complete absence of salt domes and halokinetic tectonics : nothing has disturbed the original nature and arrangement of the different facies (apart from diagenetic phenomena) nor the distribution of thicknesses (apart from solution effects and inevitable compaction).

Furthermore these layers have been explored by an extraordinarily dense network of exploratory oil drillings. Most of these boreholes show a relatively full range of logs : usually electric log and microlog or microlatorolog, caliper, gamma-ray, neutron, sonic and density logs.

Analysis of these logs permits a precise definition of the succession of main lithological units, and, frequently, an explanation of stratigraphic correlations may be posed and the zones may be identified quite rapidly for further detailed analysis from core samples. These cores themselves are also particularly numerous and are available for any scientific study (as are logs) ; finally they are remarkably stocked, in particular at the *Core Storage center of the Conservation Board of Alberta,* in Calgary (Alberta, Canada).

Such a large evaporitic basin, giving so many data and occasionally of economic interest (Rainbow and Zama gas fields , potash mines in Saskatchewan, etc.), has, of course, been subject to much study, which has allowed an approach to most of the problems commonly set by large evaporitic basins on carbonate platform. We may note some of these problems.

1. The first one concerns the carbonate buildings : their extent, their reefal or non-reefal nature. Time and space relationships between carbonate buildings and surrounding evaporites are also worth noting : to know whether the carbonate buildings entirely precede the surrounding evaporites or if there are more or less transitionnal passages between them and thus less contemporaneity.

Other main problems are :

2. The depositional depth of evaporites which bury these carbonate constructions.

3. The segregational processes of different salts in a geographical scale and the mechanisms involved.

4. The genesis of finely laminated carbonate layers (possibly interpreted as stromatolite then as planktonic sedimentation under two water bodies).

5. The interpretation of minutely folded carbonate facies, or domed ones (stromatolites, "slump", etc.).

6. Finally, the possibility of more or less partial drainage of the basin, resulting in a lowering of the water level in all the zone above a reefal barrier.

The three drill holes partly shown in the following plates are geographically located on the map in Fig. 11. All three of them are in northeastern Alberta and each of them offers different successions in the Keg River formations as well as in the overlying Muskeg formation. The studied area is part of the basin still quite close to the carbonate barrier and to marine facies in the North-West. This means that a large portion of the Muskeg formation may still be made up of dolomites and anhydrite (and not at all of rock salt or potash) and it is mainly dolomitic and anhydritic facies which will be illustrated in the following plates. One of the three boreholes, however, contains a sizeable saliferous succession in the Muskeg formation which is featured in two plates. Since this book concentrates on evaporitic formations, the Keg River limestones will not be the main subject of these plates. But in one of the boreholes, they will be studied, for here, there is a remarkable vertical succession from (1) benthic limestones – facies normally present on platforms – to (2) finely laminated, restricted, euxinic and typically pre-evaporitic limestones.

In December 1976, the accompanying plates had already been prepared for a long time, but the key had not then been composed. And now, the loss of my sight has made it difficult for me to compose an exact commentary for them. In fact, I would never have been able to without the very efficient help I received from André Maurin, Denise Noël and Jean-Pierre Prian. To them, I would like to express my gratitude and especially to André Maurin, whose personal commentary to my own pictures renders this work his almost as much as mine. . .

It would like to thank the Canadian Organisations who helped my work there, particularly the *Geological Survey of Canada* in Calgary, Alberta ; the *Core Storage Center of the Conservation Board of Alberta* (Calgary) ; the Oil Companies : *Total North America, Elf Canada, Shell and Imperial.*

I would also like to thank the technicians of the *Laboratoire de Géologie du Muséum National d'Histoire Naturelle* (Paris), especially P. Gaulier, Destarac, G. Baucher, M. Sevestre and N. Day.

SELECTED REFERENCES

Busson (G.), 1968, "La sédimentation des évaporites. Comparaison des données sahariennes à quelques théories, hypothèses et observations classiques ou nouvelles". *Mém. Mus. Nat. Hist. Nat.*, Paris, (n.s.), C, **19**, 3, 125-169.

Busson (G.), 1972, "Principes, méthodes et résultats d'une étude stratigraphique du Mézoïque saharien". *Mém. Mus. Nat. Hist. Nat.*, Paris, (n.s.), C, **19**, 44 p.

Busson (G.) and Noël (D.), 1972, "Sur la constitution et la genèse de divers sédiments finement feuilletés ("laminites"), à alternances de calcaire et de matière organique ou argileuse". *C.R. Acad. Sci. Fr.*, **274**, 3172-3175.

Davies (G.R.) and Ludlam (S.D.), 1973, "Origin of laminated and graded sediments, Middle Devonian of Western Canada". *Bull. geol. Soc. Amer.*, **84**, 3527-3546.

Dumestre (A.), "Relations entre hydrocarbures et environnement évaporitique à Rainbow Alberta (Canada)". *Rev. Ass. Fr. Techniciens Pétrole,* Paris , **194**, 29-46.

Fuller (J.G.C.) and Porter (J.W.), 1959, "Evaporites and Carbonates : two Devonian basins of Western Canada". *Bull. Canad. Petrol. Geol.*, **17** , 2, 182-193.

Holter (M.E.), 1969, "The middle Devonian Prairie Evaporite of Saskatchewan." *Sask. Depart. of Miner. Resources,* **123.**

Klingspor (A.M.), 1969, "Middle Devonian Muskeg Evaporites of Western Canada". *Bull. Am. Ass. Petrol. Geol.*, **53**, 4, 927-948.

Maiklem (W.R.), 1971, "Evaporite drawdown : a mechanism for water-level lowering and diagenesis in the Elk point basin". *Bull. Canad. Petrol. Geol.*, **19**, 2, 487-503.

Mc Camis (T.G.) and Griffith (L.S.), 1967, "Middle Devonian facies relationships, Zama area, Alberta". *Bull. Canad. Petrol. Geol.*, **15**, 434-467.

Mc Crossman (R.) and Glaister (R.P.), 1964, *Geological History of Western Canada.* Alberta Soc. *Petrol. Geol.*, **53**, 4, 927-948.

Shearman (D.J.) and Fuller (J.G.C.), 1969, "Anhydrite diagenesis, calcitisation and organic laminites, Winnipegosis formation, middle Devonian, Saskatchewan". *Bull. Canad. Petrol. Geol.*, **17**, 4, 496-525.

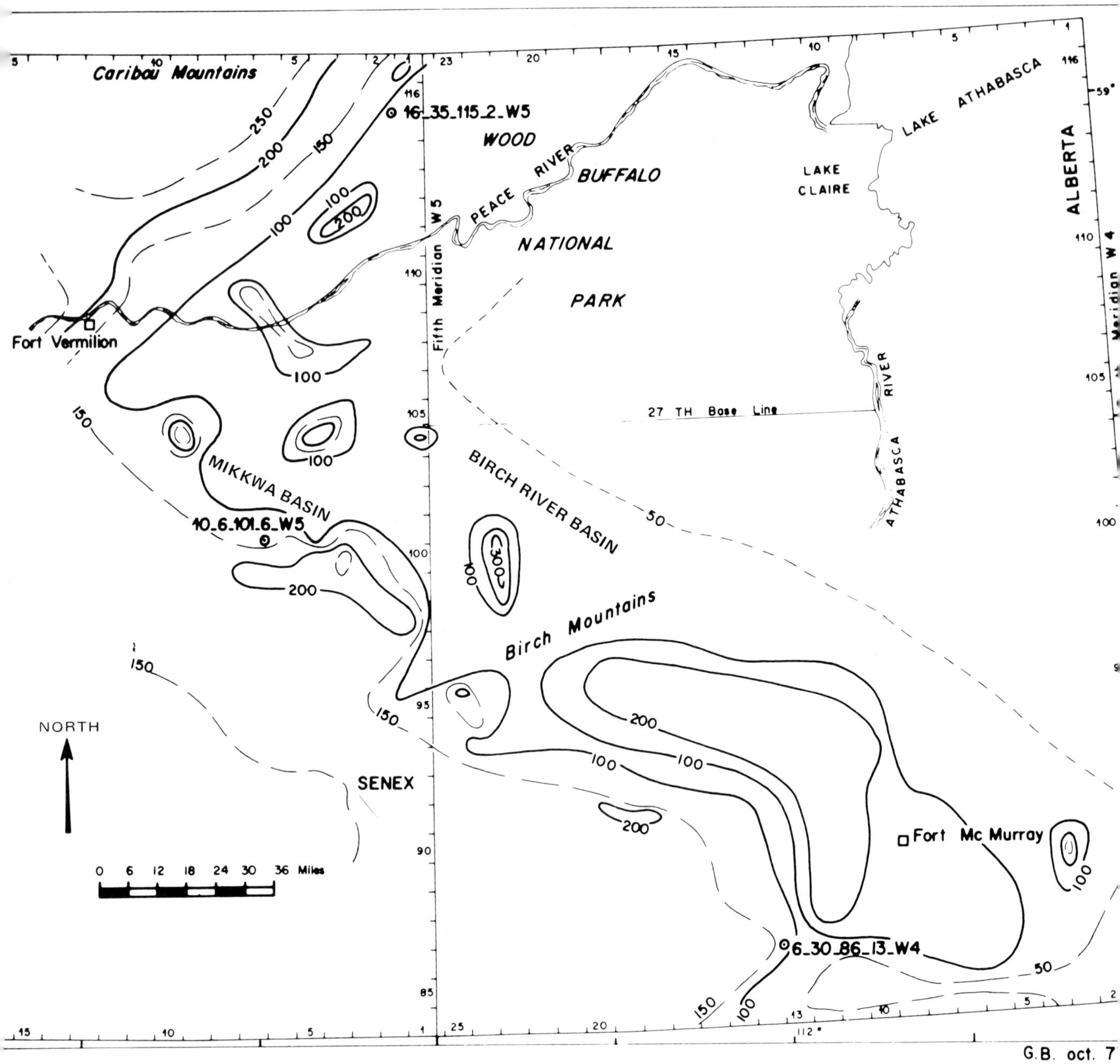

Fig. 11. – Isopach map of the Keg River formation.
Thickness (in feet) of limestone in the Keg River formation, underlying the evaporitic Muskeg formation in northeastern Alberta. The width of this limestone proves to be of considerable importance in determining the thickness and type of evaporites in the overlying Muskeg.

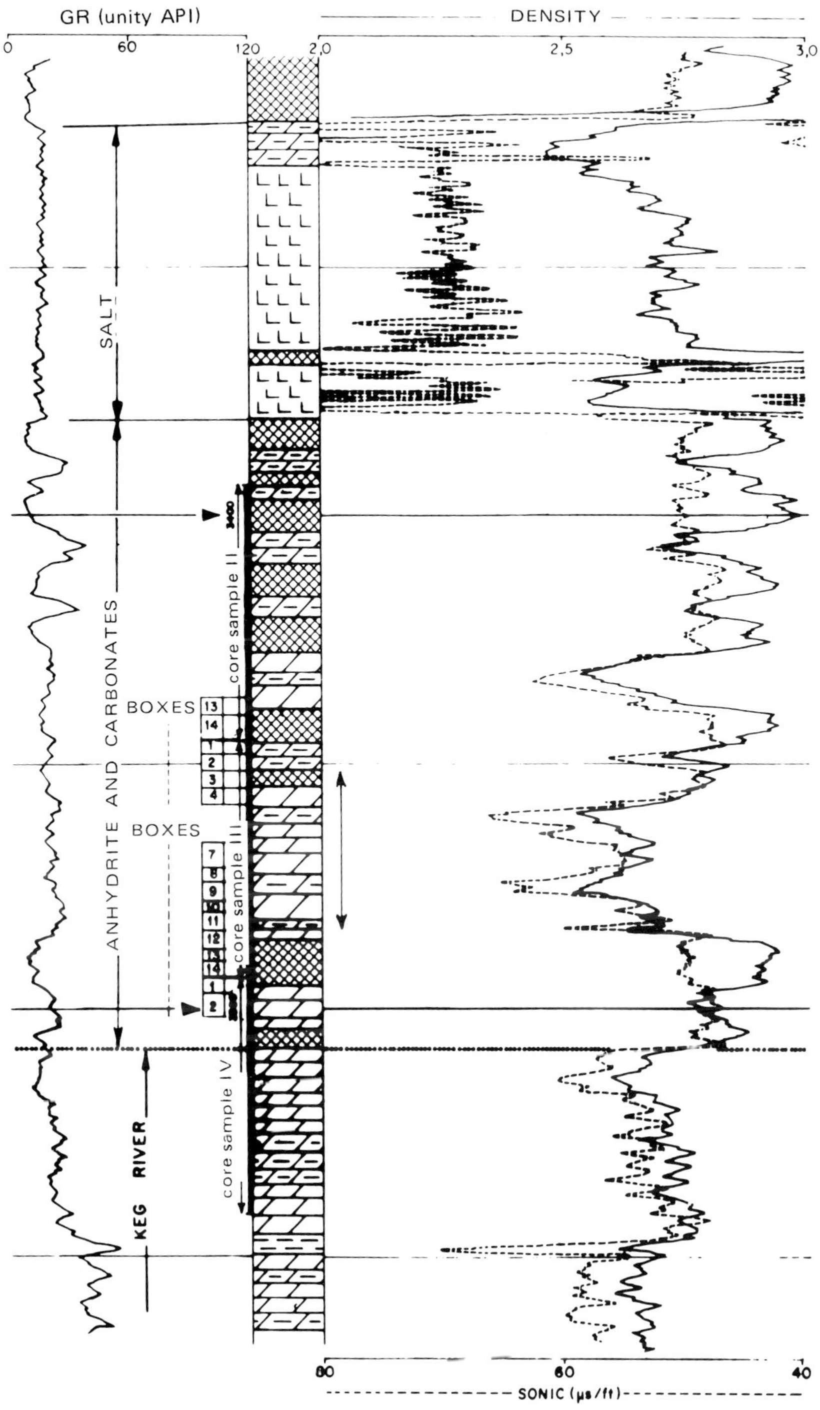

Fig. 12. – Logs of Keg River and Muskeg from borehole 16-35-115-2W5.

The Keg River limestones are 30 m thick (94 ft) and surmounted by carbonate-anhydrite alternations designated by the presence of the first evaporitic interbedding as part of the Muskeg formation. Only this system of alternations, 40 m thick (128 ft), will be described here, particularly its lower half. From the lithological profile, this system of alternations seems to be followed by rock salt. The logs shown in the adjoining plate (gamma-ray, density and sonic) provide a precise lithological determination (lithological nature and limits). We shall tentatively mention that these recordings show cyclical phenomena, important for an understanding of sedimentary mechanisms. These cycles seem, according to the gamma-ray log, to result from either a relative discontinuity in the amount of clay sediments or, as is more likely, relative irregularities in the organic matter retained in these layers. The densi-logs or sonic logs propose more complex possibilities. Take for example, the interval 3486 ft-3456 ft (marked by an arrow on the figure) in which the density curve shows recurrence of three main deflections, each one bearing a small median "crater". This type of curve form may be used for a sedimentological interpretation. A. Maurin and M. Riguidel are undertaking a study of mathematical morphology, aiming to provide an interpretation of these repeated phenomena.

V

SEQUENTIAL STUDY OF THE TOP OF THE EVAPORITIC SERIES OF THE LOWER LIAS IN A WELL IN THE AQUITAINE BASIN (AUCH 1) SOUTH-WESTERN FRANCE

J. BOUROULLEC *

Plates 31 to 37

We shall illustrate the main facies observed at the top of this series, by photographs of polished and thin sections (Plates 31-37).

These facies can be put in order following the schematic sequence below, this sequence itself forming part of the megasequence, Fig. 13.

SCHEMATIC SEQUENCE OF THE UPPER PART OF THE EVAPORITIC SERIES OF THE LIAS IN AQUITAINE
(Average thickness : 4 m, variance : 2 to 9 m)

Facies 6	Anhydrite	Chicken-wire structure
Facies 5		Enterolithic structure
Facies 4	Dolomicrite	Brecciated
Facies 3		With anhydrite nodules
Facies 2		With pseudomorphoses of gypsum by calcite
Facies 1		Bedded, bird's eyes ? pellets ? algeous encrustings

These facies and their ordering show some striking similarities with those of the sebkha type region, either present day (Shearman, 1966) or of the past (Bosellini and Hardie, 1973).

(*) Département Géologie, Société Nationale Elf Aquitaine (Production), Centre Micoulau, 64001 Pau, France.

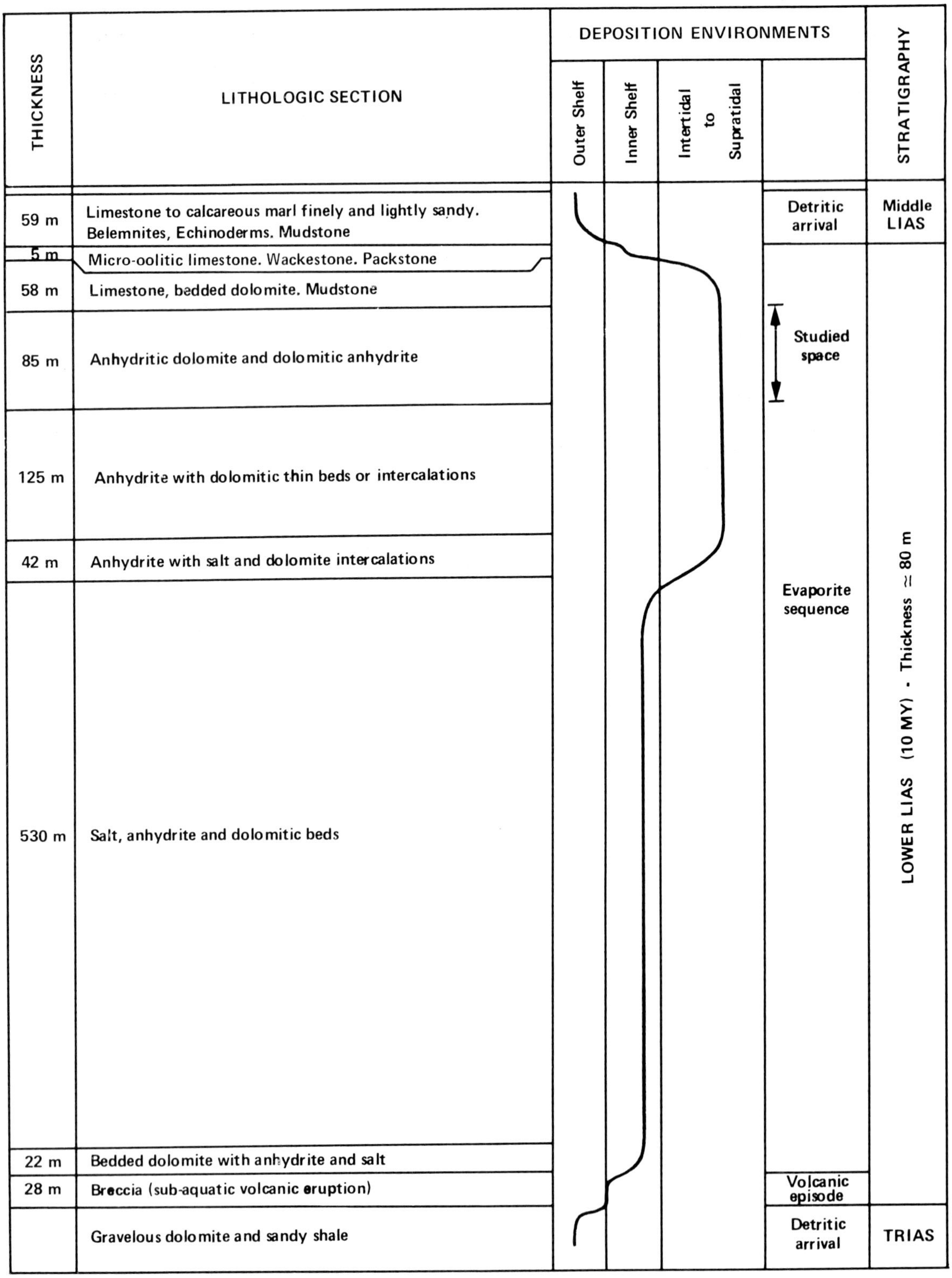

Fig. 13. – Complete evaporite sequence of the Lower Lias in Auch 1.

This possible interpretation fits in with the geological knowledge of the region : filling of the basin at the upper part of the Lower Lias, submerged laminites from the end of the cycle. Some care must be taken however: were the conditions of deposition leading to the similarity of facies exactly the same in the past as they are today ?

SELECTED REFERENCES

Bosellini (A.) and Hardie (L.A.), 1973, "Depositional theme of a marginal marine evaporite". *Sedimentology,* **20**, 5-27.

Friedman (G.M.), 1972, "Significance of Red Sea in problem of evaporites and basinal limestones". *Am. Ass. Petrol. Geol. Bull.,* **56**, 1072-1086.

Fuller J.G.C.M. and Porter (J.W.), 1969, "Evaporites and carbonates ; two Devonian basins of western Canada". *Bull. Canad. Petroleum Geol.,* **17**, 2, 182-193.

Kinsman (D.J.J.), 1969, "Modes of formation, sedimentary associations, and diagnostic features of shallow-water and supratidal evaporites". *Am. Ass. Petrol. Geol.* Bull., **53**, 830-840.

Shearmann (D.J.), 1966, "Origin of marine evaporites by diagenesis". Trans. Instn. Min. Metall., *B. Appl. Earth Sci.,* **75**, 717, B208-B215.

Shearmann (D.J.) and Fuller (J.G.), 1969, "Anhydrite diagenesis, calcitization, and organic laminites, Winnipegosis formation, Middle Devonian, Saskatchewan". *Bull. Canad. Petroleum Geol.,* **17**, 4, 496-525.

Triat (J.M.) and Truc (G.), 1974, Evaporites paléogènes du domaine rhodanien". *Geogr. Phys. Geol.* dyn., **2**, 16, 2, 235-262.

VI

THE EVAPORITIC SEQUENCES OF THE TERMINAL MIOCENE OF SICILY AND OF SOUTHERN SPAIN

J.M. ROUCHY *

Plates 38 to 44

In the terminal Miocene, the whole Mediterranean basin is particularly characterized by the individualization of a strong evaporitic episode, extremely developed at the level of the present-day abyssal plains (Joides drilling campaigns, Leg 13 and 42 A), and which outcrops at the present time, in a certain number of marginal basins on the edge of the Mediterranean, notably in Sicily, and in southern Spain. Whilst regional studies reveal interesting variations concerning the distribution and the vertical succession of the different terms of the series, the detailed observation of the main types of gypsum, of their sedimentary structures, and of their internal relations with the other constituents, reveal the same genetic processes.

1. TERMINAL EVAPORITIC MIOCENE IN SICILY

1.1. Regional framework.

The most extensive of the Miocene evaporitic basins in Sicily (the Caltanissetta basin) limited to the north by the Sicani and Madonies mountains, is situated in the central part of the island (Fig. 14). It is precisely in this basin that the most complete sequence of all the perimediterranean basins of this age appears, ranging from gypsum to potassic salts (Ogniben, 1957, Décima and Wezel, 1971, Seheiber, 1974).

The evaporitic series is separated from the white pelagic marls (Trubi facies) of the Lower Pliocene, by a detritic horizon (arrenazzolo) ; it has been subdivided by A. Decima and F.C. Wezel into two superposed unities (Fig. 14), limited by a discordance but its mutual relations are in reality masked by important tectonic deformations.

(*) Laboratoire Géologie, Museum National d'Histoire Naturelle, 43 rue de Buffon, 75005 Paris, France.

4

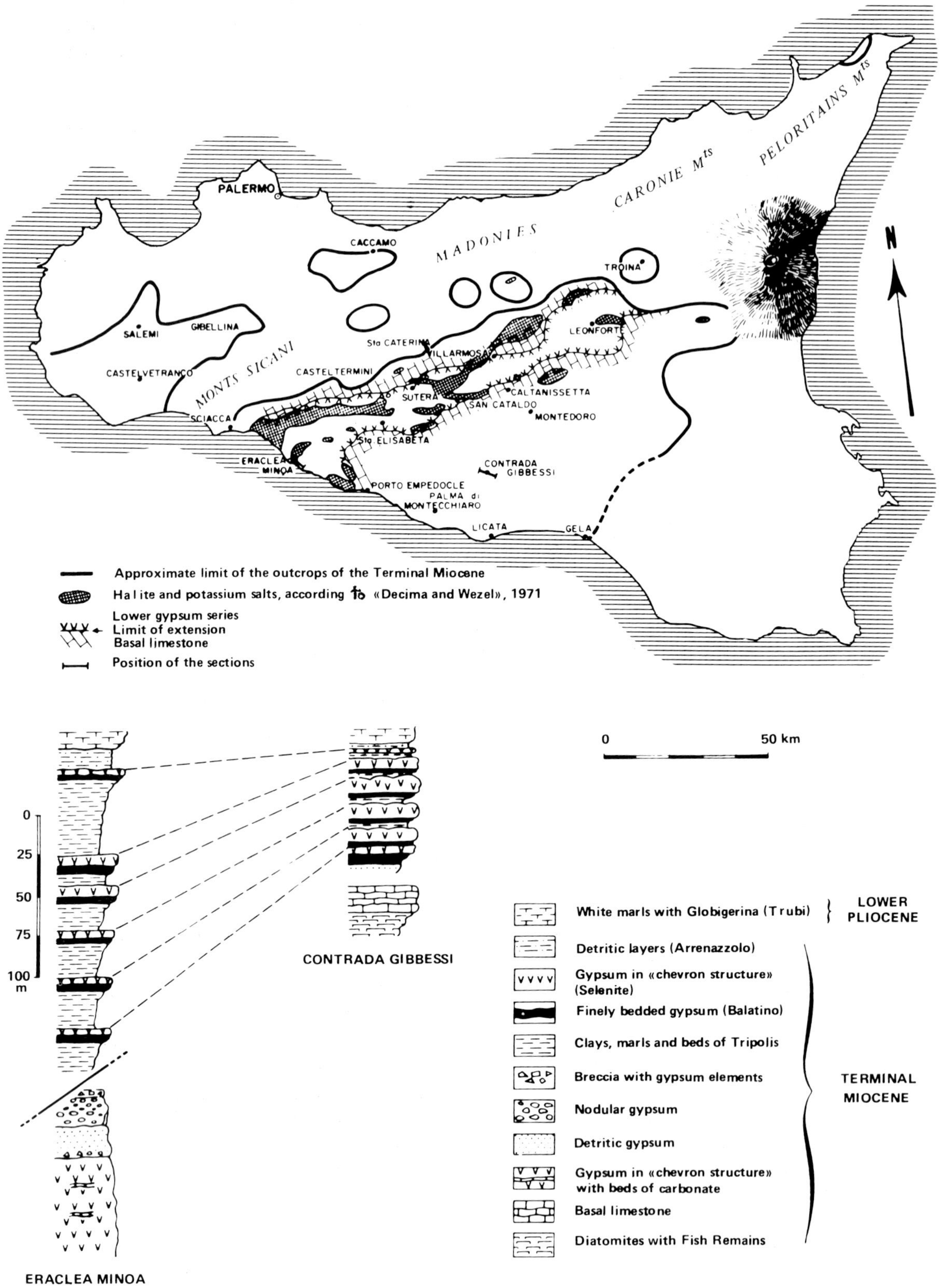

Fig. 14. – Location map and sections of the terminal evaporitic Miocene of Sicily.

The massive and strong "lower" series is localized in a narrow axial band of the basin and is composed essentially of "chevron"-gypsums or of nodular gypsums associated with carbonates. According to the authors, it is in the terminal part of this series that the very developed saliferous episode occurred (halite and potassium salts).

The "upper" series generally comprises six superposed sequences, each composed, from the bottom upwards, of : clays, marls and dolomitic carbonate laminites, finely bedded gypsum (balatino), associated with detritic gypsums and "chevron"-gypsums (selenite). This series which extends very largely beyond the limits of extension of the preceding one, evolves laterally, from the centre towards the edges of the basin, by progressive disappearance of the argillaceous intervals of each sequence, very developed in the axial part of the lower series ; the selenitic beds, which are the only ones represented on the edges, can hence be correlated with almost the whole of the basin.

On the edges of the basin, this upper series therefore covers one or more fairly thin limestone beds, which Decima and Wezel consider as being the lateral equivalents of the lower series and which are themselves superposed on a powerful series of tripolis with Fish remains.

1.2. The upper series : sequences, different types of gypsum, sedimentary structures and their relations with the surrounding sediments.

Plate 38 (Fig. 1) illustrates a classic sequence (previously described) of reduced thickness, characteristic of the edges of the basin (Palma di Montecchiaro). The finely bedded or balatino gypsum (Plate 38, Figs. 3 and 4) composed of alternating beds of gypsum, 2 or 3 mm in thickness and of limestone laminae or inframillimetric pelitics, with bedding plane or contorted, always forms the base of the selenetic banks. These beds are composed of large size crystals, twinned according to the plane (100) and orientated perpendicular to the bedding, with the point always directed towards the bottom (Plate 38, Fig. 2). Each balatino term encloses an abundant limestone nannoplankton, thus indicating sedimentation in a sea water level. In each gypsum bed, an important morphological degeneration of this nannoplankton is linked to the ecological deteriorations which occur when the concentration of the environment allows the precipitation of gypsum (Rouchy, 1975).

The existence in the balatino itself, of sedimentary structures : ripple-marks (Plate 39, Fig. 1), synsedimentary slidings towards the central part of the basin (Plate 39, Figs. 2 and 3), reworkings *in situ* of fragments of gypsum bedded before the deposition of next level (Plate 39, Fig. 4) also illustrate its primary sedimental nature. Beds of detritic gypsums frequently inserted in the finely bedded gypsum, also present crossing stratifications (Plate 39, Fig. 5), rill-marks, etc.

The selenitic banks show internal crystalline arrangements in fans diverging towards the top of the bank (Plate 40, Fig. 1) causing the formation of large size domes (Plate 40, Fig. 2), sometimes crossed by a system of radial slits : the saw-teeth aspect (Plate 38, Fig. 1 ; Plate 40, Fig. 1) of the lower limit of these banks is due to the pointed part of each fan, the crystallization of which leads to considerable buckling of the underlying balatino. Nodular gypsum frequently appears in this series under varied facies : isolated nodules in a carbonate (Plate 41, Fig. 2) or gypseous (Plate 41, Fig. 4) matrix, bedded aspect with enterolithic contortions (Plate 41, Fig. 2), mosaic texture (Plate 41, Fig. 3).

1.3. Massive gypsum of the lower series.

This series is characterized by the association selenite or nodular gypsum and dolomitic carbonates (Plate 42, Fig. 1) ; the thickness of the sequences of a different type from those of the upper series varies by several decametres from the dimension of the sample (Plate 42, Fig. 2) ; they begin by a layer of massive selenite gypsum, progressively enriched towards the top in carbonates which are sometimes finely laminated, up to individualization of a fairly thin carbonate bank. The enclosing gypso-carbonate sediments undergo a plastic deformation at the crystal contact (Plate 42, Fig. 3) and the carbonate grains are captured in the gypsum crystals which mould their contours (observation with electronic scanning microscope : Plate 42, Fig. 4).

2. EVAPORITIC TERMINAL MIOCENE IN SOUTHERN SPAIN

Evaporitic basins in southern Spain show modest dimensions comprising more reduced series without saliferous formations ; they can be divided into two types: (Rouchy, in press) the internal series, proximal to the saliferous occidental Mediterranean basin (Murcie, Sorbas, Carboneras, Almeria) and the distal external series (Abanilla, Alcantarilla, Alhama de Murcia, Lorca, Grenada ; see Fig. 15).

The evolution of these basins shows a fundamental difference which has affected the evaporite's diagenetic history: whereas, after evaporite deposit, the marine conditions persist, at least in the central part of the internal basins (Montenant, 1973), a final infilling of the marine basin and the passage to continental series takes place in the external basins, immediately after evaporite deposition.

2.1. The internal series.

Making a thicker formation than in the external basins, the beds of gypsum (selenite partially transformed into nodular gypsum) alternate locally with layers of marls with planktonic foraminifera (San Miguel de Salinas) (Rouchy, in press). They become individualized in the central part of the basin and pass towards the edges and the high points, with neritic limestones and algal and reefal constructions. They cover a powerful pelagic marl series. The tripolis are fairly thin and localized in a sector without evaporites (Murcia Basin).

2.2. The external series.

An exclusively nodular or bedded gypseous episode of minor thickness (30 m maximum) appears at varied stratigraphic levels, from the Upper Tortonian to the terminal Miocene, in a series of depressions orientated NNE-SSW between Abanilla and Grenada (Fig. 15).

The series of the Grenada Basin differs from the preceding ones by the greater thickness of the evaporitic episode, by its localization at the top of the turbiditic sequences which are sometimes coarse (with dark laminites intermediary) and finally by the presence of selenite gypsum in the western part of the basin. At Lorca, the tripolis with Fish remains are very developed at the base of the gypsum.

2.3. Some sedimentological aspects.

The gypseous intervals of the internal series are almost exclusively composed of selenite gypsum (total absence of balatino, gypsarenites and salt) in large crystals (Plate 43, Fig. 1). The analysis of their relations with enclosing sediments (Plate 43, Fig. 4) shows that there is a marine primary crystallisation in the muddy sediment. The

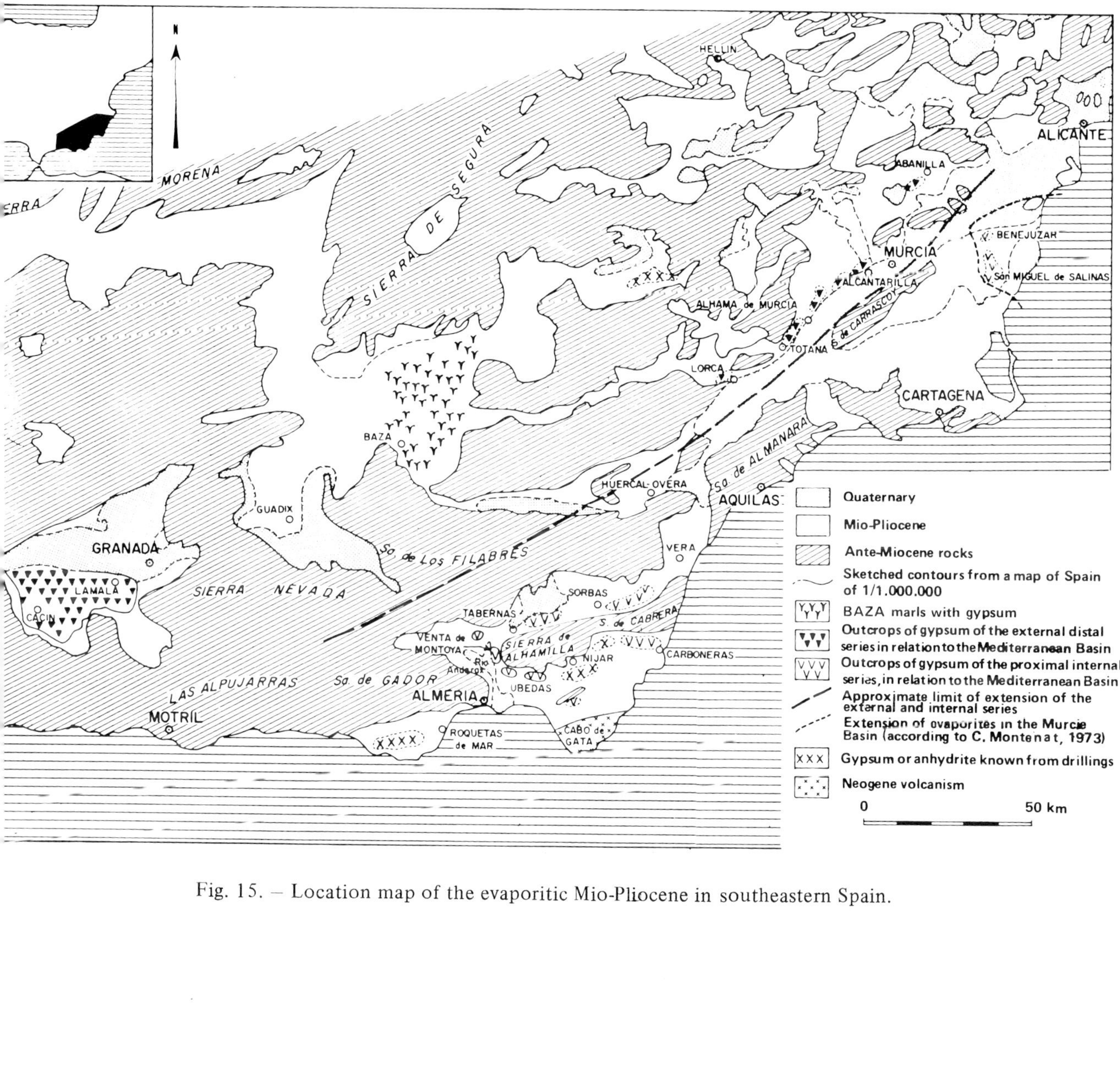

Fig. 15. – Location map of the evaporitic Mio-Pliocene in southeastern Spain.

crystallisations near the water-sediment interface are also very common. The nodular gypsum of mosaic texture (Plate 43, Fig. 3) represents a product of transformation of the selenite crystals.

The bedded nodular gypsum, its localization towards the centre of the basin and the transition to massive anhydrite nearer the centre seem to indicate the action of phreatic surfaces during burial.

These external series are composed almost exclusively of bedded gypsum showing sedimentary structures which demonstrate marine character (sole marks, Plate 44, Fig. 1) and in which nodular gypsum is growing according to an extremely complex process (Plate 44, Figs. 2 and 3). This nodular gypsum, which sometimes has an anhydritic precursor, will represent early diagenetic phenomena occurring in the marine bedded gypsum, after the infilling of the basin (the same phenomenon can occur in some supratidal sebkhas of the Persian Gulf).

3. CONCLUSIONS

The example of the evaporitic basins chosen on opposite edges of the evaporitic Miocene domain of the western Mediterranean, illustrates certain common features of evaporitic sedimentation, despite the important regional differences related to the specific paleogeographical characteristics of each region, which have affected the evolutions and general distribution of the facies. The Sicilian evaporitic series which is more complete than that of southern Spain, shows numerous similarities with that of the deep parts of the Mediterranean (existence of two superposed series including a lower saliferous episode).

The deposit of evaporites shows numerous common features ; the principal types of gypsum (thin laminated or selenite) represent primary marine crystallisations in a muddy sediment (very early diagenesis) or at the sediment-water interface. The nodular gypsum formation takes place differently in the two considered cases : relation with phreatic level during burial or superficial diagenetic phenomena after the basin infilling.

SELECTED REFERENCES

Decima (A.) and Wezel (F.C.), 1971, "Osservazioni sulle evaporiti messiniane della Sicilia centro meridionale". *Riv. Min. Sic.*, **22**, 130-134, 15 p.

Montenat (C.), 1973, "Les formations néogènes et quaternaires du levant espagnol (Provinces d'Alicante et de Murcie", *Thèse Doct.*, Paris-Orsay, 1 170 p.

Ogniben (L.), 1957, "La pétrographia della seria Solfifera siciliana e considerazioni geologiche relative". *Mem. Desa Carta geol. Ital.*, Roma, **33**, 1, 275 p.

Rouchy (J.M.), 1975, "Mise en évidence de nannoplancton calcaire dans certains types de gypse finement lité (balatino) du Miocène terminal de Sicile et conséquences sur la genèse des évaporites de cet âge". *C.R. Acad. Sci.*, Paris, Sér. D, **282**, 13-16.

Rouchy (J.M.) (in press), "Sur la genèse de deux types principaux de gypse (finement lité et en chevrons) du Miocène terminal de Sicile et d'Espagne méridionale". *Rev. Géogr. phys. et Géol. dyn.*, à paraître.

Schreiber (B.C.), 1974, "Upper Miocene (Messinian) evaporite deposits of the Meriterranean basin and their depositional environments". *Thèse,* New-York, 1974.

VII

PETROGRAPHICAL DATA OF OLIGOCENE EVAPORITES IN THE BRESSE AND VALENCE BASINS (EASTERN FRANCE)

Y. GUILLEVIN *

Plates 45 to 64

Many studies have been carried out on the Oligocene in the Bresse and Valence area, but these studies often deal only with outcrops, on the edge of the basin. The wells do, however, give good quality material, invaluable for petrographical studies and for analysis of the nannoflora and nannofauna contained. This petrographical study was facilitated by detailed lithological analyses carried out by geologists of *Institut Français du Pétrole,* such as Mr Pinel (Valence) and Mr Mathieu (Bresse). The samples and documents used were supplied by the Ground Storage Service of *Gaz de France* under the directorship of Mr Blondin, to whom I am very grateful for his close collaboration.

1. GEOGRAPHICAL AND STRUCTURAL SETTING

1.1. The Oligocene basin of the Bresse (Fig. 16)

The Bresse trough with a length about ten times its width (100 km long and 10 km wide), is drained by the Saône and its tributaries. It marks the boundary between two geologically different areas : to the east, the Subalpine mountains and the Jura, to the west, the Massif Central. The northern limit is represented by the rise of the Morvano-Vosgien Mesozoic shoal (plateau de Langres) and the Serre crystalline massif, while the southern limit has been fixed by geophysical studies, as between Lyon and Crêmieux. The trough was formed at the beginning of the Tertiary. From sudden contact between the surrounding mountains and the filling sediment, the Bresse basin would be thought to have a rift-like structure, and this is proved by the presence of N-S faults on the edge of the Massif Central, and others NE-SW along

(*) Laboratoire de Géologie, Muséum National d'Histoire Naturelle, 43 rue de Buffon, 75005 Paris, France. ESSO REP, 213 cours Victor Hugo, 33321 Bègles, France.

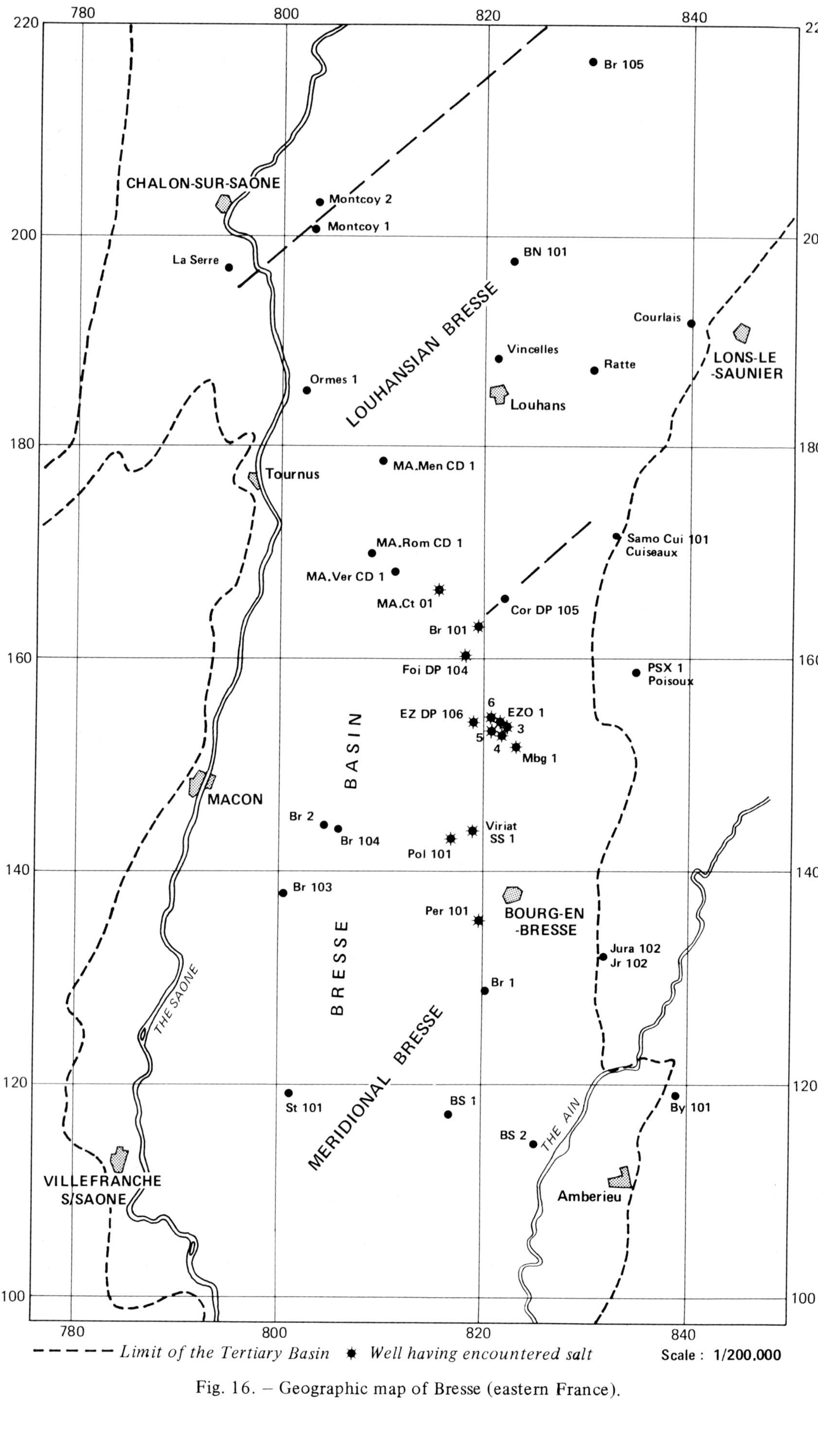

Fig. 16. – Geographic map of Bresse (eastern France).

the Jura mountains. Telluric and gravimetric maps show many shoals of Variscan direction (NE-SW) which may have affected depositional conditions of Oligocene evaporites. These shoals divide the basin into secondary regional depressions : north of the Bresse Chalonaise, which is separated by the Sennecy-La-Serre axis from the Louhansian Bresse, and, in the south, the meridional Bresse, separated from the centre of the basin by the shoal of Cormoz. The main structural features of the depression should therefore be assigned to a pre-Tertiary tectonic. Recent organic faulting affected the sedimentary strata and the general structure of the whole basin.

General lithology

The numerous wells show an unexpectedly thick Oligocene series (1 500 m at Bresse No 1). It is made up of two thick saliferous masses separated by an intermediate series with many insoluble residues. The Upper Oligocene is composed of anhydritic deposits, marls and limestones. In most of the wells, only the lower saliferous layer has been cored. The deeper evaporitic strata are separated from the shoal by a thin marly bed.

1.2. The Oligocene trough of Valence (Fig. 17)

The Valence trough running NNE to SSW is about 50 km long and 15 km wide. Its history seems to resemble that of the Bressan trough but differing in the complex pre-Tertiary tectonic stage.

Through seismic studies the limits of evaporitic deposits were established. The Beaurepaire shoal forms the northern limit to the basin ; a gravity anomaly shows a transversal "horst" shoal, from WSW to ENE. The west of the basin is bordered by sizeable faults which separate the Massif Central from Tertiary sediments. Similarly at the eastern boundary there is a series of faults parallel to the depression. To the south, the Valence trough is limited by a perisynclinal termination from the Upper Cretacean.

As in the Bresse basin, Oligocene evaporites are deposited in a depression showing warping zones which acted as barriers to reduce the force of the water and confined the area.

General lithology

Although the substratum has never yet been reached, the lower salt bed's thickness may be evaluated as approximately 500 m. A bed of dolomitic limestone, bordered by clay and anhydrite layers, separates the salt bed into two roughly equal parts. A thick detritic layer, predominantly clay facies, divides the lower salt series from a second, thinner evaporitic episode (100 m). The upper part of the Oligocene is made up of clay-sand deposits.

2. PRINCIPAL MINERALS AND CHARACTERISTIC STRUCTURES

2.1. Halite and its relations with insoluble residues

The rock salt here can take two different forms : in the Valence basin various samples show horizontal layers of insoluble matter (clays, dolomite, anhydrite) alternating with those of salt (Plate 45, Fig. 1 ; Plate 46). The halite crystals are of average size, taking on a greyish colour when containing clays and anhydrite. But in the Bresse basin, most of the salt is composed of large translucent overlapping crystals, separated by clay and anhydrite veins (Plate 46). This is, in fact, a recrystallisation salt ; the patches of intercrystalline insoluble matter were originally deposited in horizontal

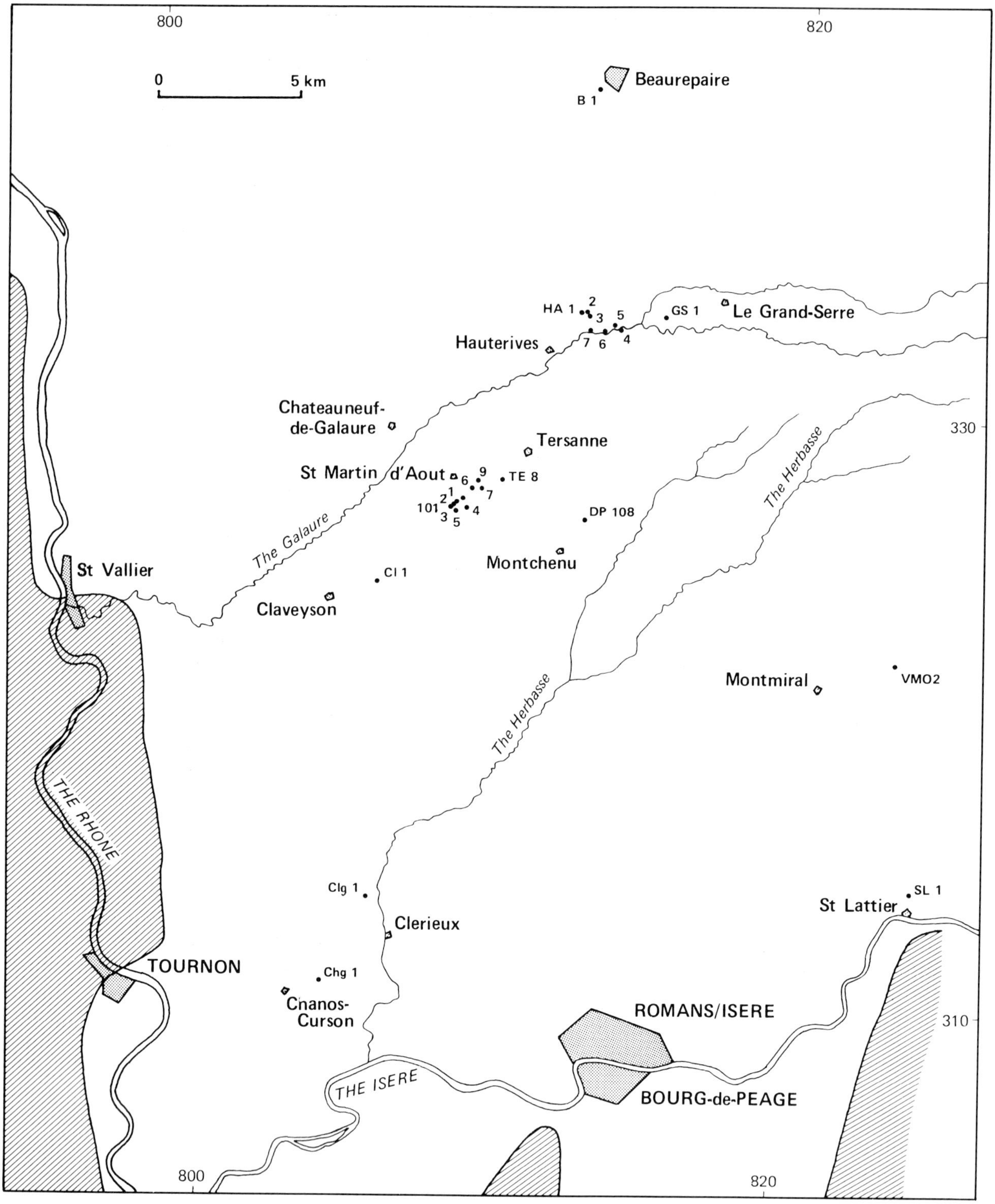

Fig. 17. – Geographic map of the Valence basin (southeastern France).

layers which show different detritic influxes in the sedimentation basin. During recrystallisation, the growth of halite crystals completely altered the primary structure of the sediment. In thin section, halite is seen to include such minerals as anhydrite – in acicular crystals – frequently associated with calcite rhombohedrons. Anhydrite is composed of a felted mass of fine needles, or lengthened phenocrystals, which, in recrystallisation halite, emphasize the boundaries of the salt cubes (Plate 48, Figs. 1 to 4). During its growth, the halite pushed the impurities contained within to the edges of the new crystals. It is difficult to identify with certainty the diagenetic stage of the minerals enclosed within the salt. We may simply note that the frequent association of anhydrite and calcite seem to show that acicular crystals result from recrystallisation of calcium from calcite, sulphates originating from brines in residual solutions.

2.2. Anhydrite : structure, texture, origin

The most usual form is nodular anhydrite (Plate 49, Fig. 1). The nodules vary in size and are joined in a salt or carbonate clay groundmass. Through an optical microscope, the nodules are seen to be made up of acicular crystals, sometimes with preferential parallel or subparallel orientation relative to bedding. Such a structure is referred to as a "gneissic structure" by some authors (Plate 49, Fig. 2). This general orientation may be modified by an obstruction such as an impurity – causing the anhydrite laths to change direction. Further on, we will see that the structure of nodules evolves during diagenesis.

Various hypotheses have been proposed to explain their formation. In 1949 Imlay suggested deposition of a gelatinous mass of calcium sulphate at the same time as that of the underlying sediment. The irregular, more or less rounded shape of nodules resulted from wave and current action on them before they were completely solidified. According to Dunham, some nodules were formed through the disintegration of massive anhydrite (Plate 50, Figs. 1, 2) or from carbonates in the ground mass.

In Bresse and Valence evaporites, most of the anhydrite nodules result from pseudomorphism of gypsum (Plate 51, Fig. 1). This means that the anhydrite nodule retains, at least at the beginning of its transformation, the crystalline form of the original mineral. This replacement can be made directly or by gypsum dissolution (Plate 51, Figs. 2, 3). Thermodynamically, this transformation is accompanied by water-loss and a decrease in volume, causing the nodule to shrink. The water released increases the surrounding sediment's plasticity which could cause distortion during growth of anhydrite crystals (Plate 52, Fig. 1), but compaction is also partly responsible for these sedimentary forms.

This "gneissic structure" may be altered during diagenesis. Thus, through insufficient place during growth of anhydrite, some nodules have folded forms, or enterolithic structure (Plate 52, Fig. 2 ; Plate 61). Acicular crystals which were parallel to bedding become perpendicular or oblique on the nodules' outer edges (Plate 53, Fig. 1). It seems that anhydritic patches, where acicular crystals have no form, originate from the association of various nodules with a complex folded structure. The three stages, dehydration, expansion and compaction, seem to have occurred before the sediment was consolidated. However, these observations are not sufficient to allow the diagenetic stage of these nodules to be established. Drilling showed that induration of sediments could not occur until Secondary period. But, where gypsum undergoes first dissolution and then recrystallisation (Plate 51, Figs. 2, 3), the transformation seems relatively early, since a cavity corresponding to crystalline gypsum forms and, partially occupied by some anhydrite crystals, is unlikely to remain unchanged during compaction. From the stage of isolated nodule, through the enterolithic structure and up to massive bed, anhydrite takes different forms, such as "chicken-wire" structure (Plate 49, Fig. 1), mosaic (Plate 53, Fig. 2), laminated (Plate 54, Figs. 1, 2) or even mammillary (Plate 55).

2.3. Silica

This can be deposited in various forms : as megaquartz, either close (in cavity filling) (Plate 56, Figs. 1, 2) or separated by a carbonate-clay groundmass, or associated with feldspars and micas (Plate 56, Fig. 3), as microquartz with located sizeable patches of extinction (Plate 57, Fig. 3), or as fibro-radial chalcedony (Plate 57, Figs. 1, 2). The latter form shows negative elongation and, according to Folk, this mineralogical peculiarity arises through polymeration of silice, caused by a lowering in the environment's pH (disappearance of calcium sulphate). Elsewhere, silica is found closely associated with anhydrite (Plate 59, Fig. 2). Chalcedony layers several millimetres thick progress vertically to massive anhydrite. A mixture of the two minerals lies between the two layers. Could this perhaps be due to replacement of anhydrite by silica? Nodular structure ghosts are found in the massive chalcedony (Plate 56, Figs. 1, 2). This silica may therefore originate from a relatively late epigenetic stage, the anhydrite resulting from gypsum transformation. Progressive replacement of macro-crystalline anhydrite by silica is also evident in certain examples of cavity-filling (Plate 58, Fig. 1).

2.4. Laminites

Laminite sediments result from alternating carbonate layers and organically rich clays. The proportions of these two elements vary according to the environment, giving a sediment predominantly light or dark (Plate 59, Figs. 1, 2). In the Valence trough, laminites within the evaporitic series contain abundant Coccolithophoridea nannofauna relative to the number of specimens (Plate 60, Fig. 1). The coccoliths in dolomitic carbonate layers (Plate 62, Fig. 2) are perfectly conserved, very small (1 μ) and belong, on the whole, to just one species of the *Reticulofenestra* group which, stratigraphically, could only appear in the Paleogene, a fact which rules out the theory of later reworking.

A carbonate bed at the top of the main salt mass contained normal size coccoliths, such as *Braarudosphaera* (Plate 60, Fig. 3), associated with *Reticulofenestra.*

Laminites and their contents give us an idea of their depositional conditions, as well as those of evaporites. The suggestion of a marine origin of laminated deposits is confirmed by the presence of coccoliths. However, the reduced size and monospecificity of these fossils suggest a confined environment and the complete absence of benthic organisms confirms this. The euphotic zone containing the coccoliths was probably already oversalted. The numerous coccospheres suggest deposition during calm conditions. The limestone beds, rich in coccolith species, above the salt layer, emphasize the progressive dominance of marine influences over regressive evaporitic conditions (Plate 60, Fig. 4).

3. EVAPORITIC CYCLES

In a sedimentary evaporitic cycle, the deposition of different sulphates and chlorides was ordered according to their solubility and density levels : in an ideal cyclothem, we would see the following order : bituminous clay, limestone, anhydrite, halite, anhydrite, limestone, organically rich clay. To complete the cycle, we would have to include the deposits of sodium chloride after those of potassium and magnesium salts. From first studies, they seem absent from the Bresse and Valence basins, either through their feeble concentration, or sedimentary hiatus and return to the sea, or through post-depositional erosion. I have chosen a sample from each basin to illustrate the formation of a minor, centimetric, evaporitic cycle.

3.1. The Bresse basin (Plate 61)

Green to beige marls containing anhydrite lenses and some rock-salt crystals. A fissure vein of red fibrous salt runs through the whole sample. The evaporitic cycle is separated from the lower part which is seen, according to the description of the core, to be made up of beige marls containing some anhydrite lenses. The upper part of the sample of green laminated clay beds, marks the start of another sedimentary cycle.

At a certain concentration of brine, calcium sulphate is deposited in various forms:

(a) Small nodules in horizontal layers, sometimes grouped together.

(b) Larger nodules.

(c) Anhydrite with enterolithic structure (in this case, the structure originates from the coalescence of anhydrite nodules).

Sulphate deposition can be interrupted by simply diluting the brine. Laminites mark the end of the evaporitic cycle, and the start of the following one.

3.2. The Valence basin (Plate 62 to 64)

At the top of the evaporitic series, the regressive chloride sedimentation gives way to layers of insoluble matter. Next to the salt layers is a deposit of sulphate in the form of massive anhydrite (Plate 64, Fig. 1), overbedded by laminites, with prevailing carbonates (No 1). Organic matter becomes progressively more abundant (No 2). The end of the first transgressive sedimentary cycle is marked by carbonated clays. The following cycle starts with a lenticular gypsum layer (No. 3) which is dissolved and partially recrystallised into anhydrite (Plate 63, Figs. 1, 2). The deposit of sulphate continues as layered (No. 4) in a carbonate clay groundmass (Plate 64, Fig. 2). The end of this incomplete regressive cycle is marked by mosaic anhydrite. Evaporation could not have been strong enough for the brines to reach the saturation level when sodium chloride starts to precipitate. Several centimetres higher in the series, another transgressive cycle begins with new carbonate deposits (Plate 64, Fig. 1).

SELECTED REFERENCES

Busson (G.) and Noel (D.), 1972, "Sur la constitution et la genèse de divers sédiments finement feuilletés (laminites) à alternance de calcaires et de matière organique ou argileuse." *C.R. Acad. Sc.*, Paris, **274**, 3 172-3 175.

Delwig (L.F.), 1955, "Origin of the Salina salt of Michigan." *J. Sed. Petrol.*, **25**, 2, 83-110.

Folk (R.L.) and Pittman (J.S.), 1971, "Length-slow Chalcedony ; a new testament for vanished evaporites." *J. Sed. Petrol.*, **41**, *4*, 1045-1058.

Friedman (G.M.), Amiel (A.J.), Braun (M.) and Miller (D.S.), 1973, "Generation of carbonate particles and laminites in algal mats.Example from sea, marginal hypersaline pool, Gulf of Aqaba, Red Sea." *Am. Ass. Petrol. Geol. Bull.*, **57**, 3, 541-557.

Duff (S.) and Thomson (A.), 1963, "Origin of nodular and bedded anhydrite in Permian shelf sediments, Texas and New Mexico." *Am. Ass. Petrol. Geol. Bull.*, **47**, 9, 1726-1732.

Murray (B.C.), 1964, "Origin and diagenesis of gypsum and anhydrite." *J. Sed. Petrol.*, **34**, 3, 512-523.

Posnjak (E.), 1940, "Deposition of calcium sulfate from sea water." *Am. Journ. Sci.*, 238, p. 559.

Wardlaw (N.C.) and Schwerdtner (W.M.) 1966, "Halite-anhydrite seasonal layers in the Middle Devonian prairie, evaporite formation, Saskatchewan, Canada." *Bull. Am. Geol. Soc.*, 77, 331-342.

Withington (C.F.), 1961, Origin of mottled structure in bedded calcium sulfate", *Geological Survey Research,* 410.

third part

DIAGENESIS AND EVAPORITIC DEPOSIT SEQUENCES

VIII

DIAGENETIC ANHYDRITE OF CARBONATE SEBKHA IN THE TERTIARY OF IRAQ

R. CUSSEY *

Plates 65 to 69

The origin of evaporites has been the subject of much controversy. This seems to arise from the relative scarcity of present-day environments of saliniferous series deposits.

The study of the sebkhas was concentrated in particular on the coast line near Abu Dhabi and allowed the role of early diagenesis to be assessed in the laying down of these evaporites. This deposition is related to evaporation under arid climatic conditions in the supratidal zone at the level of the sediment's capillary fringe.

The analysis of present day evaporitic deposits and those of the past (in this case, the Tertiary series of Iraq) permits the many convergences in facies to be re-united and accords some evaporites, at least in the most favorable cases, to a shallow coastal origin.

1. ANALYSIS OF THE EVAPORITIC SERIES IN THE TERTIARY OF IRAQ

The Oligocene series studied corresponds to the upper part of the Asmari formation. It is basically carbonate with subsidiary clay sands and evaporites. The evaporitic part in the general sedimentary evolution corresponds to confined marine deposits.

1.1. The characteristic elementary order (within the series studied).

Examination of sediments showed that anhydrite was distributed in series of deposits 20′ to 50′ (5 to 15 m) thick, in the following lithological sequence (Fig. 18 and Plate 65) represented schematically and reading from top to bottom :

(*) Laboratoire de Géologie, Société Nationale Elf Aquitaine (Production), Centre de Boussens, 31360 St-Martory, France.

Facies	Depositional environment
Dolomicritic mudstone	Supratidal
Dolomitic wackestone with scarce Foraminifera and Ostracodes	High intertidal
Algal dolomitised boundstone, brecciated, bioturbed or with solution facies Algal dolomitised boundstone Pelletoidal dolomitised grainstone	Intertidal
Dolomitic wackestone – packstone with benthic oligospecific Foraminifera	Subtidal

It is a lagoon series with particularly well-developed algal facies.

1.2. Morphological features of anhydrite.

Anhydrite is distributed in various different forms within this sequence :

(a) At the top, it is represented mainly as abundant micropatches (Plate 66, Fig. 1) ; in this upper part, irregularly shaped centimetric nodules (in chicken mesh) are sporadic and often replaced by dolomite (Plate 66, Fig. 2).

(b) Lower down, (at 15′6″), the nodular structure takes predominance but the lower limit of each concentration is clearly marked, below which the nodules disappear completely (Plate 67, Fig. 1).

(c) Towards the bottom, (at 33′), the nodules disappear and anhydrite becomes pseudo massive and irregularly shaped, but the lower limit is also clearly defined (Plate 67, Fig. 2).

Analysis using thin sections shows that anhydrite micropatches may show varied structures :

(a) In the upper part, they appear as monocrystals and may occur either from replacement of a dolomite zone (Plate 68, Figs. 1 and 2) or from filling of a dissolution vacuole (Plate 68, Fig. 3).

(b) In the lower part, micropatches have a different morphology : they result from a filling of dissolved bioclastic elements or of integranulary spaces by a complex system of lengthened, tangled crystals (Plate 69, Fig. 2). Nodular anhydrite in chicken meshes or chicken wire and massive anhydrite are found, however, solely in a complex multicrystalline form (Plate 69, Fig. 1).

2. INTERPRETATION OF OBSERVED FACTS. THE ORIGIN OF ANHYDRITE IN THIS SERIES

In the presented case, anhydrite is proved, therefore, to be of early diagenetic origin ; it results :

(a) Either from replacement of dolomite (with many ghosts of elements, especially bioclastic ones).

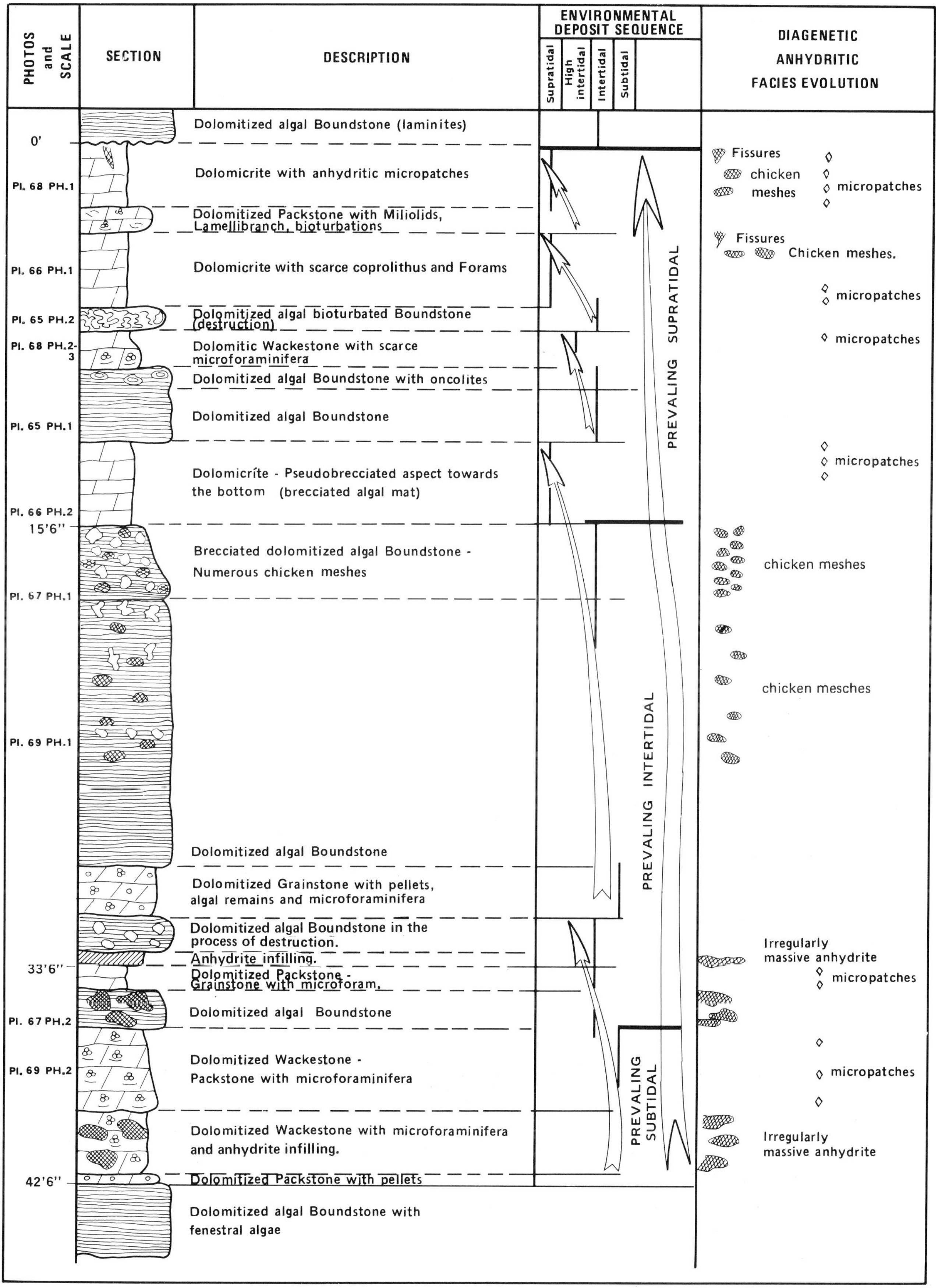

Fig. 18. – Lagoonal deposit sequence and associated anhydritic forms.

(b) Or from refilling of dissolution vacuoles.

(c) Or from refilling of dissolved organisms.

(d) Or from mineral growth (nodules) into a fine sediment which has undergone plastic deformation, showing its early origin.

Note : A later anhydrite was also observed in the refilling of fissures.

This anhydrite may be assigned to a sebkha supratidal emersion phase. Such a hypothesis is based on :

(a) The sedimentary sequence : with the abundance of algae (algal-mat), some marked confined deposits, the well developed supratidal zone.

(b) The early secondary origin of anhydrite confirmed by the plastic deformation of the surrounding sediment, the variety of anhydritisation forms, relations with the depositional sequence, the sharp interruption at the base of the concentrations (phreatic water table) and its likeness to common studied recent structures.

(c) Various demonstrations of diagenetic action in fresh water, such a vadose vacuolary dissolution, dissolution of dissolved organisms (phreatic phase) and the presence of a drusy mosaic (phreatic water).

The different types of anhydrite are not distributed in any one way, they show an evolution of vertical sequence which is probably linked to a supratidal phase at the end of the deposition in elemental order.

Anhydrite shapes	Crystallisation	Depositional evolution
Replacement anhydrite micropatches Vacuole refilling anhydrite micropatches	Each patch with single crystal	Top
± dolomitised chicken-meshes Abundant chicken-meshes Irregular massive anhydrite Refilling of dissolved elements anhydrite micropatches	Patches of confused crystals	Bottom

This vertical distribution of phenomena in the succession of deposits (5 to 15 m thick) seems to contradict the suggestion of successive water levels, which is otherwise compatible with their actual observation (0.5 to 1.50 m thick maximum).

There are, in fact, two distribution orders of anhydrite :

(a) The first agrees with the existence of successive water levels produced by ruptures and concentrations, which would explain the complexity of the distribution of morphological types.

(b) In the other, there seems to be an interdependence between the depositional environment (which itself confirms the extent and duration of emersion) and the dominance of an identified morphological structure.

At the very base, probably through lack of emersion, dissolution of elements prevails over anhydritic concentrations which do not occur. Later the spaces will be filled with anhydrite.

When emersion phases occur, the anhydrite nodules develop at the level of the successive water levels, and their coalescence may result in anhydritic masses.

In the upper part, which is nearly always emergent, the concentration becomes too weak to cause regular production of nodules, so they become sporadic, and vadose vacuolary dissolution occurs. Anhydritic formation will be restricted to the partial filling of vacuoles and replacement.

3. CONCLUSIONS

In this text, we have presented a series of sedimentary structures and diagenetic forms which confirm the frequency of sebkha sedimentation in former evaporitic formations.

The Tertiary series studied includes numerous sequential pulsations, each corresponding to a filling phase. Within these series, various diagenetic sequences of anhydrite develop.

The link between the two types of sequence is not always clear, being obscured by various early and late phases of diagenesis.

SELECTED REFERENCES

Bush (P.R.), 1970, "Chloride rich brines from Sebkha sediments and their possible role in ore formation". Trans. Instn. Metall., *B. Appl. Earth Sci.*, **79**, 137-144.

Buttler (G.P.), 1969, "Modern evaporite deposition and geochemistry of coexisting brines, the Sebkha, Trucial Coast, Arabian Gulf". *J. Sediment. Petrol.*, **39**, 1, 70-89.

Butler (G.P.), 1970), "Holocene gypsum and anhydrite of the Abu Dhabi Sebkha, Trucial Coast : an alternative explanation of origin". *Third north. Ohio Geol. Soc. Salt Symposium,* (Ed. by Rau (J.L.) and Dellwig (L.F.)), Cleveland, Ohio, **1**, 120-152.

Butler (G.P.), Kendall (C.G.St.C.), Kinsman (D.J.J.), Shearman (D.J.) and Skipwith (Sir P.A. d'E.), 1965, "Recent evaporite deposits along the Trucial Coast of The Arabian Gulf" *Proc. Geol. Soc. London.,* **1623**, 246-252.

Evans (G.) and Shearman (D.J.), 1964 , "Recent celestine from the sediments of the Trucial Coast of the Persian Gulf". *Nature,* Lond., **202**, 4930, 385-386.

Kendall (C.G.St.) and Skipwith (Sir P.A. d'E.), 1969, "Holocene shallow-water carbonate and evaporite sediments of Khor al Bazam, Abu Dhabi, Southwest Persian Gulf". *Bull. Am. Ass. Petrol. Geol.,* **53**, 841-869.

Kinsman (D.J.J.), 1969, "Modes of formation, sedimentary associations, and diagnostic features of shallow-water and supratidal evaporites". *Bull. Am. Ass. Petrol. Geol.*, **53**, 830-840.

Purser (B.H.), 1973, *The Persian Gulf*, Springer-Verlag, New-York, 471 p.

Shearman (D.J.), 1966, "Origin of marine evaporites by diagenesis". Trans. Instn. Min. Metall., *B. Appl. Earth Sci.,* **75**, 717, B 208-B 215.

IX

HABITATS AND DIAGENESIS OF MIOCENE GYPSUM BENEJUZAR (ALICANTE) SPAIN

A.F. MAURIN *

Plates 70 to 72

The reader is referred to C. Montenat's thesis of 1974 : "Les formations néogènes et quaternaires du levant Espagnol", *Thesis,* Orsay for a stratigraphic description of the outcrop.

Slightly further south in the San Miguel de Salinas zone, previously studied by Montenat, detailed petro-sedimentalogical studies have been carried out by F. Orti Cabo and D.J. Shearman : "Upper Miocene Gypsum : San Miguel de Salinas, South-eastern Spain" (in press).

(*) Compagnie Française des Pétroles, 39-43 quai André-Citroën, 75739 Paris Cedex 15, France.

X

EVAPORITIC HABITATS OF THE MICHIGAN SILURIAN (USA)

P. BOUSQUET, A.F. MAURIN *

Plates 73 and 74

The sedimentological model used by geologists of the *Compagnie Française des Pétroles* differs little from that of *Amoco* published by Mesollela *et al* (1).

The reader can refer to these authors for the names of evaporitic or carbonate structures, as well as for the relations between reefs and their surrounding evaporites.

(1) Reference : Mesollela (K.J.), Robinson (J.D.), McCormick (L.M.) and Ormiston (A.R.), 1974, "Cyclic Deposition of Silurian Carbonates and Evaporites in Michigan Basin". *Bull. A.A.P.G.*, vol. **58**, N° 1, pp. 34-62.

(*) Compagnie Française des Pétroles, 39-43 quai André-Citroën, 75739 Paris Cedex 15 et Laboratoire Central, Compagnie Française des Pétroles, 114 cours Galliéni, 34404 Talence, France.

XI

EVAPORITES IN A SUBSIDENT CONTINENTAL BASIN (LUDIAN AND STAMPIAN OF MORMOIRON-PERNES IN SOUTHEASTERN FRANCE) SEQUENTIAL ASPECTS OF DEPOSITION PRIMARY FACIES AND THEIR DIAGENETIC EVOLUTION

G. TRUC *

Plates 75 to 87

INTRODUCTION : THE PALEOGENE EVAPORITIC BASINS OF SOUTH-EASTERN FRANCE : GENESIS, STRUCTURE AND ORIGIN OF MATERIALS

The Mormoiron-Pernes Paleogene basin forms part of a complex of continental sedimentary basins, distributed along the whole length of the Rhône-Saône axis in the southeast of France. From north to south are found successively : The Bressan rift--valley, the Dauphinois graben, which is extended by the Crest (Drôme) basin and supported on the western spurs of Vercors, followed by the relatively unimportant Tricastin basin. The widening of the Rhône valley starting from this sector can be noted in the Paleocene, due to the existence of several subsident grabens, separated by more stable platforms (Triat and Truc, 1974). These are : the Alès and Camargue grabens, the Mormoiron-Pernes and Apt-Forcalquier basins, which present very localized points of subsidence, and finally the Manosque graben on the Durance border.

a. Genesis of the evaporitic basins.

As a general rule, all the Paleogene grabens and basins in this region contain evaporitic deposits of essentially Ludian formation. Two later and weaker episodes are also

(*) Departement des Sciences de la Terre, Université de Lyon, 27-43 boulevard du 11 Novembre 1918, 69621 Villeurbanne, France, et Laboratoire associé au CNRS n° 11, "Paléontologie stratigraphique et Paléoécologie".

recorded in the series : one in the Middle Stampian, the other in the terminal "Chattian".

In the Ludian, the simultaneous existence of grabens in rapid subsidence and of more stable platforms results from tectogenesis in an E – W extension which follows the Lutetian and Bartonian compression or so-called "Pyreneo-Provençal" phases.

In the southern part of the Rhône-Saône valley – which forms the main object of the results presented – this tectogenesis is emphasized by the faults at 30° N – 40° N, which limits the subsident central area of the basins and grabens to that of the Alpine and Central Massifs (Fig. 19) : these are the Alès and Durance faults, of which the structural role remains very important during the Paleogene.

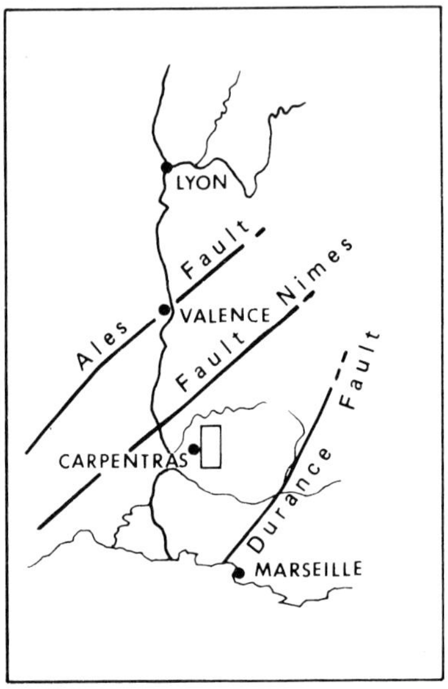

Fig. 19. – Geographic and structural location of the Paleogene evaporites in the Mormoiron region, near Carpentras (Vaucluse). This is between two families of major accidents, N 30°-N 40°, which were in distension in the Ludian.

These structural parameters determine a paleogeography which is very favorable to the formation of evaporites; indeed, in such a morphotectonic framework in extension, the basins and grabens become quickly isolated, and form a purely endoreic unity, separated by reliefs from the only marine domain known at this time in the southeast of France, i.e. the nummulitic furrow of the subalpine (secondary) chain.

It is not certain that the genesis of these structural confinements could have led to the formation of the above-mentioned evaporitic complex without the periods of aridification of the environment. Indeed, within the same structural framework, detritic, fluviatile carbonate or lacustrine deposits exist in the Stampian, but without any evaporitic traces, and the dulçaquicoles fauna of these deposits characterize an environment regularly supplied with fresh water.

b. Place and nature of the evaporites in the Paleogene series.

It must be noted that the whole Rhône-Paleogene unity is formed by a succession of detritic phases, and phases of either carbonate or evaporitic origin. In considering only the Ludian-Chattian interval, three major detritic episodes can be seen, emphasized by

			Mormoiron-Pernes	Apt-Murs	Forcalquier	Manosque	Durancian Area
Discordant Marine Miocene							
Oligocene	Chattian		Beaucet limestones with Helix ramondi	Reillanne limestones	Mourres limestones Sigonce limestones and lignites	Marls of the Manosque's tiling	Durancian coarse detritic complex
	Stampian		Vénasque limestones and marls	Viens Marls		Bois-d'Asson thick-bedded limestones formation	
				Vachères limestones		"Montfuron limestones"	
				Caseneuve marls		«Brown-sepia» marls and sandstones	
			Laminites with fishes			«Little cubes» limestones	
			Isle-sur-Sorgue limestones and gypsum	Limestones, dolomites and gypsums			
	«Sannoisian»		Valette de Pernes green sand and Unang marls	Pradenque sandy marls	Revest sandy marls	"Mort d'Imbert" upper sandy red marls	
	Lower stamp.		Limestones with Brotia laurea	La Fayette limestones	Lower detritic banks	Lower finely-bedded limestones Lower red marls Marls with Cyrena	
	Ludian	upper	Mormoiron evaporitic formation	Gargas gypsiferous shales Limestones with cyrena	gap ?	Supra saliferous marls with anhydrite (Evaporitic complexe)	
		lower	Mormoiron sandstones and green shales Methamis limestones Crillon conglo	Jean-Jean sandstones and fluviatil marls («glauconitic sandstones»)		Saliferous complex infra-saliferous marls (Evaporitic complexe)	
Eocene	Bartonian	upper	Jocas limestones Mormoiron attapulgites	Jean-Jean limestone with attapulgite	?	Sandstones and sands with lateritic gravels	
		lower	Red argillaceous sandstones with lateritic gravels				
	Lutetian	upper	(gap)	Apt and Murs limestones	Carbonated siliceous crusts sandy marls with formation microcodium	Conglomerates with microcodiums	
		lower		Sandy marls and conglomerates with microcodiums			

Fig. 20. – Biostratigraphical outline, showing the sequential position of the Ludian and Stampian evaporitic layers (Mormoiron-Pernes basin).

Such depositis develop in alternation with the detritic formations, at the time of periods of aridification. Then follow more humid periods, capable of spreading coarse material into the basin under fluviatile influence (stratigraphic denominations : geological maps at 1/50 000 Carpentras, Cavaillon, Reillanne, Forcalquier and Manosque).

vast spreadings, which precede each episode, where an evaporitic sedimentation is dominant over all the basins mentioned (1) (Fig. 20)

On a global scale, it can be affirmed that each remission of the detritic deposits represents a disequilibrum of the climate, which evolved very rapidly towards an aridification of the environment; restricted climatic and sedimentologic conditions were hence added to the restricted structural conditions, the whole leading inevitably to the genesis of evaporites. In correlation with this, the detritic materials – in relation with an uninterrupted tectonic during the whole of the Paleogene – are accumulated on the borders of the basin, without spreading, and without any possible transit towards the centre. The model obtained sometimes possesses many affinities with that of the Dead Sea – and with other subsident grabens – where the accidents on the border are fringed by a narrow band of coarse detritic materials, which pass very rapidly towards the centre of the basin to the silty argillaceous laminites, as well as the large saliferous masses (Gubler *et al.*, 1975) interbedded with black-shales.

Another significant phenomenon concerns the segregation of evaporitic material, the nature of which varies according to the sedimentary trap considered. **Halite (rock salt) and the important layers of anhydrite are always localized in the strongly subsident grabens ; on the contrary, on the platforms which separate these grabens, halite is absent, anhydrite is rare, and gypsum is largely dominant.** The best regional example is provided by the co-existance in the Ludian of the Manosque graben (Alpes de Haute-Provence), halitic and anhydritic, extended by the Forcalquier-Apt and Mormoiron platform, where a gypseous sedimentation accompanied by some banks of anhydrite develop, even in the most subsident sector (Mormoiron) (2).

The argillaceous minerals also reveal this fundamental difference ; in the Manosque graben, the argillaceous assemblage of the evaporites is formed exclusively of chlorite and illite, which derive for the most part, by diagenetic agradation, from an initial detritic material (Sittler and Triat, 1975). On the platforms, however, the magnesian smectites and the sepiolite are of neogenetic origine (Triat and Trauth, 1972 and specially by Trauth, 1974). Schematically speaking, it seems that one sole cause might be able to explain this group of observations.

The effect of subsidence and the proximity of active accidents brings about an afflux of fine particles at the level of grabens, whilst the platforms, which are further away from these sectors, benefit from more episodic detritic deposits, and tend towards an isolation which leads to neoformations of the magnesian argillaceous minerals.

The same structural parameter is undoubtedly responsible for the presence of halitic masses in the grabens: crystallogenesis of the halite may occur on the platforms, as can be observed on the surface of the present day sebkhas, but it is never followed by a stable deposition of this salt. The change back into permanent solution of the halite on the platforms causes it to migrate progressively towards the grabens, where it enriches a body of dense salt-brine, caught in a strictly endoreic structure trap.

These affirmations seem to support the arguments developed by G. Busson (1968, 1974), as regards the almost exclusive presence of halitic masses in the most subsident traps, where the confinement reaches a very high degree, and where the concentration of the source-water on a fairly thick slice of the aquatic body, seems to be a fundamental condition of the genesis of these deposits.

---

(1) The last of these, very weak in the regional Chattian, is better expressed more to the south : gypsum at Aix-en-Provence (Hugueney and Truc, 1976).

(2) See the geological maps mentioned in the key to Fig. 20.

c. Origin of the evaporitic material

As regards the origin of the materials which have played a part in the genesis of Paleogene evaporites in the southeast of France, most observations tend to suggest a continental model, without any opening towards the sea : analyses of stable isotopes of carbon, oxygen, and sulphur (J.C. Fontes), presence of structural barriers between the Priabonnian sea of the subalpine chain, and the Rhône-Saône axis, possibilities of leaching of the evaporitic Trias at the outcrop, etc.

The elements of the marine nature fauna which can be found in the carbonate sediments, laterally equivalent to the evaporites, constitute one example in the geological series of something which present day nature shows us only rarely in a few areas of the globe : an adaptation of certain species, transported from a fairly close marine reservoir, to an environment in which the chemical characteristics recall, in some aspects, that of the original.

The leaching of the sedimentary formations which surround the Paleogene evaporitic basins, seem therefore to constitute the sole cause of the evaporites (Triat and Truc, 1975).

II. THE MORMOIRON-PERNES EVAPORITIC BASIN ; DYNAMICS AND EVOLUTION OF THE DEPOSITS ; STRUCTURAL AND CLIMATIC CONTROL OF SEDIMENTATION

The choice concerning the presentation of this basin was made for several reasons. The evaporitic series is known from numerous exploitations, from drillings, outcrops, and the lateral passages have not been destroyed by erosion. The evolution of the environment and the sedimentation can therefore be fully appreciated (Triat and Truc, 1972).

1. Ludian

In the Ludian, the Mormoiron-Pernes basin represents an example of a basin which shows, in some areas, the effects of a moderate subsidence favouring the concentration of evaporites without reaching the level of that of the grabens, and in particular the Manosque graben (3).

Figure 21 gives an idea of the structural arrangement of the principal point of subsidence at Mazan, as regards the paleogeographic environment, i.e. the mountain chain (already sharply modelled at this period) of the Ventoux-Baronnies region and of the Monts de Vaucluse. An intensive gypsum exploitation exists in the Malemort-Mazan sector, where the evaporites reach their greatest thickness (approximately 200 metres).

a. Dynamics of the deposition

At the end of the Bartonian, the N-S Pyreneo-Provençal phases of compression are relayed by a stretching tectonic, mentioned before. Whilst the northern margins received a strong detritic deposit (1′, Fig. 21), of which the finest materials spread out across the whole basin, (2, Fig. 21), carbonate sedimentation occurred in the east, at the

(3) These evaporites are of the same age as those of the Parisian basin, which have recently formed the object of some detailed studies (Fontes, 1974).

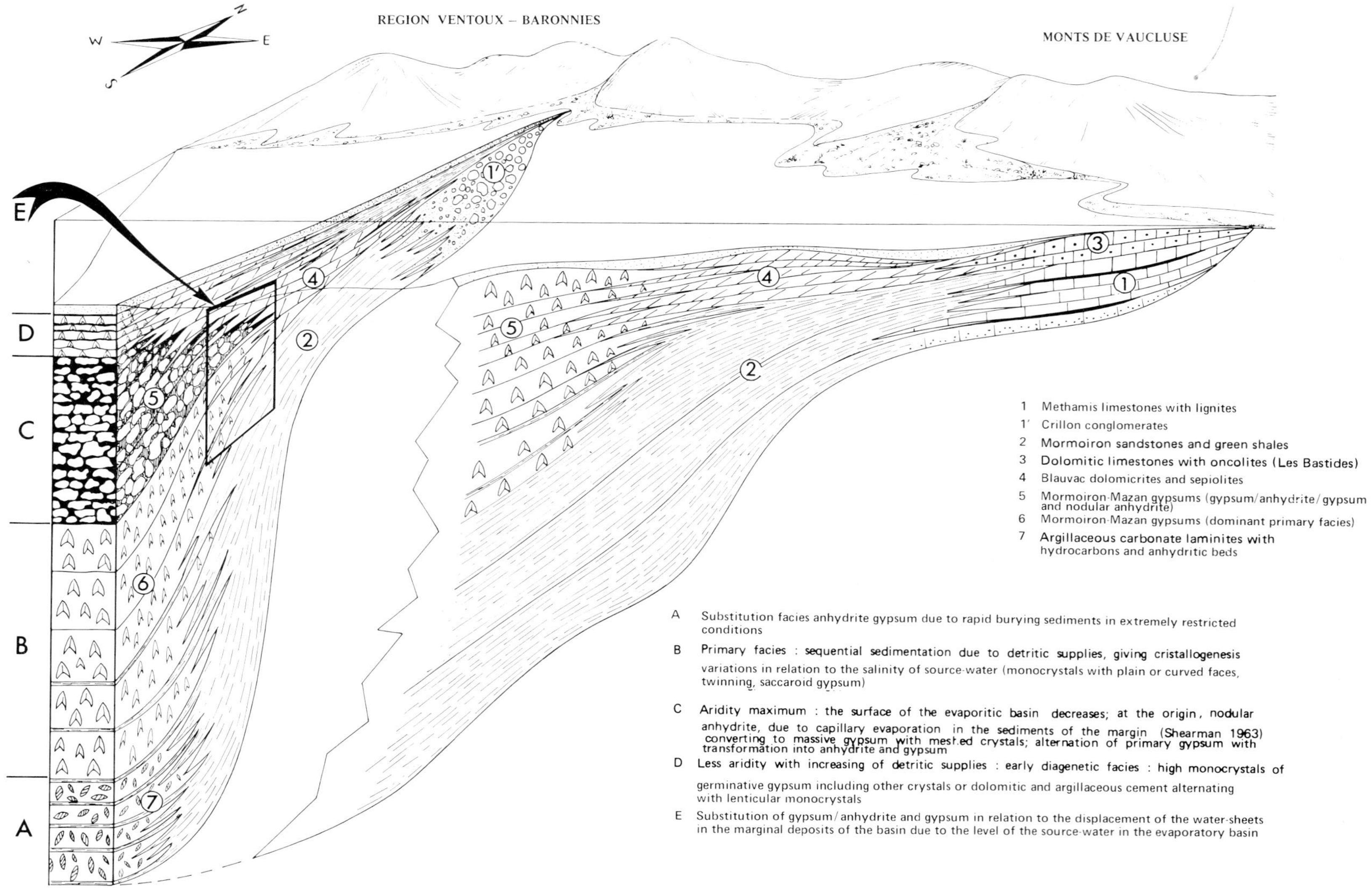

Fig. 21. – Paleographical reconstitution of the Mormoiron evaporatory basin in the Ludian.

In the Lower Ludian, the northern margins of the basin receive important detritic deposits (1 ft) ; the clays are spread out in the whole of the basin (2), whilst in the SE, a sheltered sector allows sedimentogenesis of the carbonates to occur, the fauna of which show mesohaline features. Many argilo-lignite beds are inserted in these limestones. Only the centre of the basin – already strongly subsident – sees the formation of evaporites in a very confined environment, rich in organic matter (argillaceous laminites with hydrocarbons).

In the Upper Ludian, the aridification of the environment, which is in correlation with a regression of the detritic deposits, leads to the extension of the evaporatory domain. A dolomitic aureole (4) composed of dolomicrites and magnesian smetites, separates the edge and the centre of the basin. Towards the SE, the carbonate sedimentation continues with the same faunal characteristics : dolomitic layers with oncolites and foraminifera are also included. In the centre of the basin, the facies of the gypsum deposits reflect a fluctuation in the salinity of the source-water (5,6) and in their extent. On the dolomitic borders, the variations of the level of the phreatic sheet-water leads to substitution phenomena.

point of contact with the Monts de Vaucluse (see key Fig. 21). Beginning in the Lower Ludian, a preferential subsidence affected the Mazan region, together with the accumulation of argillaceous materials, rich in organic matter in this part of the basin. The prevailing conditions of confinement at this time then allowed the first crystallogenesis of evaporites to take place. These were lenticular monocrystals of gypsum, untwinned, coming most probably from an early diagenesis, due to their oblique arrangement in the sediment (7, Fig. 21). Weak quantities of halite are associated to the black-shales facies in the deeper part of the basin. The volume of the detritic masses diminishes very progressively, whilst the climatic pulsations lead rapidly towards an aridification of the environment ; the Mormoiron basin becomes an evaporitic basin which is more and more spread out. On the basin's margin first of all, an alternance between green clay and gypsum can be noted, then, following an even clearer retreat of detritic, argillaceous minerals in a neoformation accompanied by dolomite (magnesian smectites, sepiolite and dolomicrites).

A "dolomitic belt" henceforth separates the borders of the basin and its centre, where parageneses in gypsum occur (4, Fig. 21). The sequential aspect of the deposition, the form and arrangement of the gypsum crystals, (Deicha, 1974) are determined by the water supply of the basin, which is itself subjected to numerous other factors. Amongst these, the climatic variations on diverse scales play the most important part. From time to time, the largest surface area of the basin is subjected to dewatering, a phenomenon which is recorded by the argillaceous gypsum interbanks, in the form of mudcracks covered with the prints of birds' feet and the tracks of mammals (Plate 75).

A phase of extreme aridity necessarily includes an important decrease in the water surface of the basin, and the genesis of anhydrite in nodules ("chicken-wire facies") by capilliary evaporation at the heart of the deposits (5, Fig. 21). This phenomenon, described on the Trucial coast of the Persian Gulf (Shearman, 1963, 1966 ; Kinsman, 1966) on a very minor thickness, is evident on several tens of metres in the centre of the Mormoiron basin.

The arrival of new detritic deposits marks the return to more humid climatic conditions which are favorable to an extension of the basin, and a crystallization of the primary gypsum, or of early diagenesis in the large crystals.

b. Diagenetic evolution of the deposition

Besides the early diagenesis facies linked to sequences B and D (Fig. 21, Plate 76-80) and represented elsewhere in the marine or continental evaporitic deposits there exists a diagenetic evolution of the Mormoiron and Mazan evaporites, by transformation and substitution, which brings into play the anhydrite/gypsum equilibrium in two opposite directions (Truc, 1978).

- At the level of sequence A, the rapid burying of laminites rich in organic matter (7, Fig. 21) in a reducing environment (subsident structural trap) is followed by the formation of hydrocarbons. At the same time as this process, the lenticular gypsums undergo a total dissolution, and are replaced by anhydrite, without deformation of the contour of the crystals (Plate 82). This distinctive feature excludes any possible gypsum/anhydrite transformation, which is associated with severe thermodynamic conditions (Jauzein, 1977). On the other hand, anhydrite crystallisations coming from saturated solutions impregnating the sediment and in the presence of catalysts, is much more probable.
- At the level of sequence C, a more or less complete transformation of the original anhydritic nodules in gypsum can be seen and of gypsum previously transformed into anhydrite; this is accompanied by numerous deformations of stratifications with threads/veins/filaments of fibrous gypsum at the level of the planes of friction between

the transformed banks, and the inert banks of primary gypsum (Plate 81). The aspect of the gypsum resulting from this transformation is indicative of the strong constraints which accompany the increase in volume : serrated crystals on very irregular patches, and on the metric scale of the banks, presence of large oblique cleavages on a shining surface.

Lots of residual anhydrite sometimes remain in abundance in the nodules, an observation also made on the Mediterranean Messinian evaporites (Schreiber and *al.* , 1976, p. 740).

This manner of anhydrite formation, which is rare in present day nature, seems therefore to appear frequently in geological series (Schreiber *et al., op. cit.* ; Bertrand and Jelisejeff, 1974, for the Miocene in Tunisia).

• Other facies of substitution which are less often represented, affect numerous gypseous levels ; i.e. sporadic silicifications on the level of crystals, or more generally, on the level of a bed. Plate 83 gives an idea of their nature and their situation in the Mormoiron-Pernes Paleogene evaporites.

• Finally, very particular facies of substitution (Plates 84 and 85) exist on the margins of the evaporatory basin, affecting the gypsums which alternate with the green clays or with the dolomicrites. The equilibrium between the source-water of the evaporatory basin and the phreatic sheet supplied by the slopes is very often disrupted to the benefit of the primary sources-waters which largely invade the marginal deposits. The differences in concentration between these two opposing bodies cause gypsum dissolutions-recrystallizations which may even succede the gypsum/anhydrite/gypsum transformations. Only the gypsum/anhydrite/gypsum substitution facies on beds of original lenticular monocrystals are represented. In correlation with this, the dolomicritic beds are partially replaced and this "gypsification" phenomenon may, in certain areas, attain the whole sedimental mass.

2. Stampian

The sealing of the Ludian evaporites is assured by a vast detritic spreading of the Lower Stampian (Fig. 20) and the return to arid conditions occurs in the Middle Stampian in a different context, characterized by an almost total absence of magnesium (Triat and Truc, 1974).

Stampian gypsums (Plate 86) are rarely observed under a primary facies, but tend to form in a calcitic complex where there is an alternation between ostracodes and oncoid micrites (Fig. 22). These alternations are the object of gypsum replacements under the influence of the same water-sheet phenomenon described for the Ludian margins (Fig. 21, E). The micritic cements are particularly sensitive to these substitutions, and leave behind a skeleton formed of partially eroded oncoids, in inclusion in a microcrystalline gypsum of alabaster facies (Plate 87).

The instances of replacement by gypsum of a carbonated material seem to be common, and a convergence may be mentioned between the Pernes Stampian and certain Messinian beds, noted by C. Schreiber (*et al.* , 1976, p. 743, Fig. 12).

III. CONCLUSIONS

At the end of this presentation it is difficult to propose any synthetic outline for the Paleogene evaporites of southeastern France, other than that defined by G. Busson in a recent study : ". . . when evaporites become widespread at a certain stratigraphic level of a basin, whatever the site considered (lateral sharpening, bottom of slope,

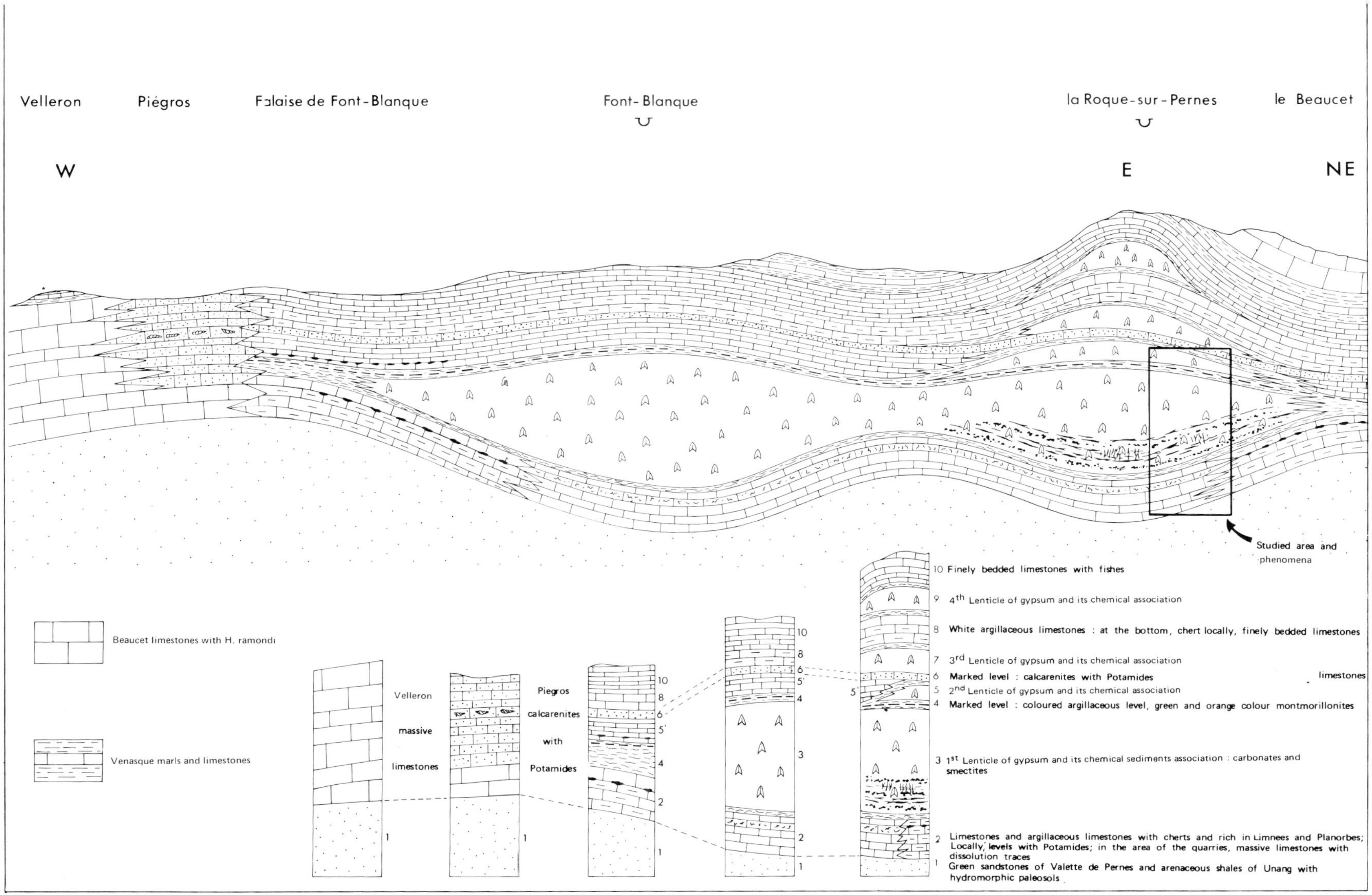

Fig. 22. – The Stampian evaporites of the Mormoiron-Pernes basin in their carbonate complex. Several lentils, elongated following the subsident axes associated with the N 40° lineaments, showing substitution facies (gypsum/carbonates). These lentils correspond to low points in the paleogeography, in which there has been circulation of the source-waters of the evaporatory basin, in relation with certain variations of depth. The sediments initially deposited (micrites with oncoids, in alternation with micritic laminites with Ostracodes) are partially or totally replaced very often by a crystalline gypsum (figured at the base of bank N° 3). Figure drawn from Triat and Truc, 1974, modified.

subsident region) it seems logical to admit that these evaporites have been deposed in conditions of different depths, from the borders into the centre of the basin. . ." (1974, p. 206).

Whether on a regional scale, in considering the Mormoiron-Manosque unity for example, or rather on the scale of a small size basin, such as that of Mormoiron, it must be stated that the evaporitic facies and their condition of genesis, are clearly very different for the same period. From the layers of nodular anhydrite on the basin's margin, the origin of which is due to processes of capillary evaporation, continuing into deep halitic masses of the Manosque graben, a very extended range of facies exist : the material of which they are composed, shows not only the initial conditions of paragenesis, in a variable structural context, but also a complete diagenetic evolution by agradation, transformation and substitution.

SELECTED REFERENCES

Bertrand (J.P.) and Jelisejeff (A.), 1974 "Formation d'évaporites par des processus d'évaporation capillaire." *Rev. Géogr. phys. Géol. dyn.*, Paris, **2**, 16, 2, 161-169.

Busson (G.), 1968, "La sédimentation des évaporites ; comparaisons des données sahariennes à quelques théories, hypothèses et observations classiques ou nouvelles." *Mém. Mus. Hist. Nat.*, Paris, N.S., C, **19**, 3, 128-167.

Busson (G.), 1974, "Sur les évaporites marines : sites actuels ou récents de dépôts d'évaporites et leur transposition dans les séries du passé". *Rev. Géogr. phys. Géol. dyn.*, Paris **2**, 16, 2, 189-207.

Deicha (G.), 1974, "Fluctuations cristallogénétiques et séquences évolutives dans le gypse parisien." *Bull. Inf. Géol. Bass. Paris,* **41**, 15-17.

Fontes (J.C.), 1974, "Le dossier du gypse parisien ; essai critique sur la notion de faciès." *Bull. Inf. Géol. Bass. Paris,* **41**, 7-14.

Gubler (Y.), Arbey (F.), Arthaud (F.), Desprairies (A.), Dubois (P.), Freytet (P.), Fontes (J.C.), Gigot (P.), Kandel (C.), Masse (P.), Masson (P.), Sittler (C.), Schuler (M.), Touray (J.C.), Triat (J.M.), Trichet (J.), Truc (G.), "Dynamique des dépôts dans un bassin sédimentaire continental ; exemple d'un bassin paléogène de Haute-Provence (Manosque, Forcalquier, Apt, Carpentras)." *9ème Congr. Intern. Sédim.*, Nice, Thème 5, 199-203.

Hollyday (D.W.), 1970, "The petrology of secondary gypsum rocks : a review." *J. Sedim. Petr.*, **40**, 2, 734-744.

Hugueney (M.) and Truc (G.), 1976, "Corrélations stratigraphiques et paléogéographie des formations marines et continentales à la limite Oligocène-Miocène dans le sud-est de la France". *Géobios,* Lyon, **9**, 3, 363-365.

Jauzein (A.), 1974, "Les données sur le système CaSO4, H_2O et leurs implications géologiques". *Rev. Géogr. phys. Géol. dyn.*, Paris, **2**, 16, 2, 151-159.

Kinsman (D.J.J.), 1966, "Gypsum and anhydrite of recent age, Trucial Coast, Persian Gulf." *Second Symposium on Salt ;* Northern Ohio Geol. Soc., Cleveland, **1**, 302-326.

Kinsman (D.J.J.), 1969, "Modes of formation ; sedimentary associations, and diagnostic features of shallow-water and supra-tidal evaporites." *Bull. Am. Ass. Petrol. Geol.*, **53**, 830-840.

Schreiber (C.), 1973, "Survey of the physical features of Messinian chemical sediments. Messinian events in the Mediterranean." *Koninkl. Nederl. Akad. Wetens.*, Amsterdam, 101-110.

Schreiber (C.), Friedman (G.M.), Decima (A.) and Schreiber (E.), 1976, "Depositional environments of the Upper Miocene (Messinian) evaporite deposits of the Sicilian basin." *Sedimentology,* Zurich, **23**, 6, 729-760.

Shearman (D.J.), 1963, "Recent anhydrite, gypsum, dolomite, halite from the coastal flats of the Arabian shore of the Persian Gulf." *Proc. Geol. Soc.*, London, **1607**, 63-64.

Shearman (D.J.), 1966, "Origin of marine evaporites by diagenesis." *Inst. Mining Metallurgy Trans.*, B, **75**, 717, 208-215.

Shearman (D.J.), Mossop (G.), Dunsmore (H.) and Martin (M.), 1972, "Origin of gypsum veins by hydraulic fractures". *Inst. Min. Metall.*, **81**, 149-155.

Sittler (C.) and Triat (J.M.) 1975, "Les argiles du fossé de Manosque ; synthèse des données et interprétations", *9ème Congr. Intern. Sédim.*, Nice, livret-guide de l'excursion A2, 44-45.

Trauth (N.), 1974, *Argiles évaporitiques dans la sédimentation carbonatée continentale tertiaire.* Thèse, Strasbourg, 309 p.

Triat (J.M.) and Trauth (N.), 1972, "Evolution des minéraux argileux dans les sédiments paléogènes du bassin de Mormoiron (Vaucluse)." *Bull. Soc. Fr. Minér. Cristallogr. phys.*, Paris, **95**, 4, 482-494.

Triat (J.M.) and Truc (G.), 1972, "L'Oligocène du bassin de Mormoiron (Vaucluse). Etude paléontologique et sédimentologique", Docum. Lab. Géol. Fac. Sci., Lyon, *Notes et mém.*, **49**, 27-52.

Triat (J.M.) and Truc (G.), 1974, "Evaporites paléogènes du domaine rhodanien." *Rev. Géogr. phys. Géol. dyn.*, Paris, **2**, 16, 2, 235-262.

Triat (J.M.) and Truc (G.), 1975, "Les évaporites paléogènes du domaine rhodanien ; un exemple de genèse d'évaporites en milieu continental." *5ème Réun. ann. Sci de la Terre,* Montpellier, 368.

Truc (G.), 1978, "Lacustrine sedimentation in an evaporitic environment : the Ludian (Palaeogene) of the Mormoiron basin, southeastern France." *Spec. Publs. int. Ass. Sediment.*, Amsterdam, 15 p.

West (I.), 1975, "Evaporites and associated sediments of the basal Purbeck Formation (Upper Jurassic) of Dorset." *Proc. Geol. Ass.*, 86, p. 205-225.

PLANCHES

PLATES

CHAPITRE I (Planches 1 à 9). *CHAPTER I (Plates 1 to 9)*

Un exemple de sédimentation évaporitique continentale actuelle. Le bassin tchadien.

An example of recent continental evaporitic sedimentation. The Chadian Basin (Africa).

PLANCHE 1 LE MILIEU NATUREL

Fig. 1 Vue aérienne des dépressions interdunaires du Kanem.

Fig. 2 Lac interdunaire en voie d'assèchement.

Fig. 3 et 4 Ramassage de la dalle de trona en fin de saison sèche.

(*Cliché* G. Maglione).

PLATE 1 NATURAL ENVIRONMENT

Fig. 1 Aerial view of interdunary depression of the Kanem.

Fig. 2 Interdunary lake drying out.

Figs. 3 and 4. Lifting of the trona slab at the end of the dry season.

(*Photographer* G. Maglione).

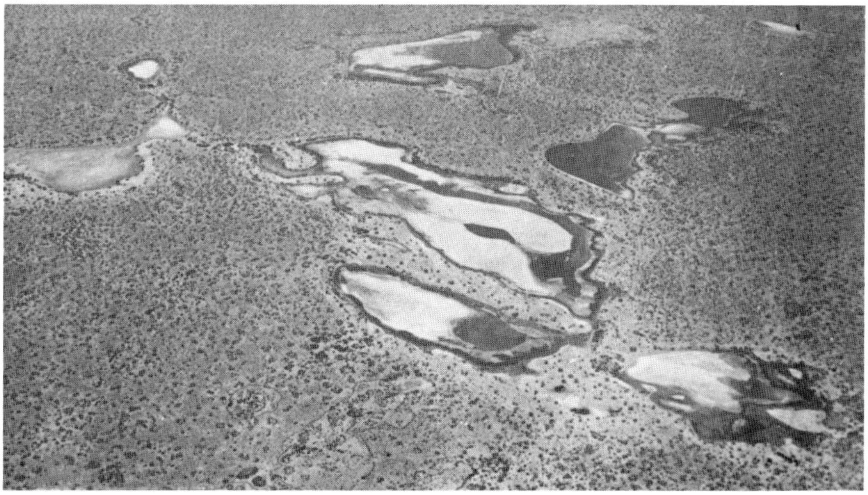

1

2

3

4

PLANCHE 2 LA PARAGENESE EVAPORITIQUE

Fig. 1 Concrétions calcaires autour de racines de "doum" ayant formé une mèche évaporante. Dégagées par déflation éolienne.

Fig. 2 Croûte saline (mélange de trona, de halite et de sable éolien) formant des "tentes de sel" décimétriques.

Fig. 3 Cristal de gaylussite.

Fig. 4 Cristal de gypse.

Fig. 5 Cristaux de natron (déshydraté en thermonatrite).

Fig. 6 Trémies de halite.

(*Cliché* G. Maglione).

PLATE 2 EVAPORITIC PARAGENESIS

Fig. 1 Limestone concretions around "doum" roots having formed an evaporating "wick". Freed by eolian deflations.

Fig. 2 Saline crust (mixture of trona, halite and eolian sand) forming decimetric "salt tepees".

Fig. 3 Gaylussite crystal.

Fig. 4 Gypsum crystal.

Fig. 5 Natron crystal (deshydrated into thermonatrite).

Fig. 6 Hoppers of halite.

(*Photographer* G. Maglione).

1

2

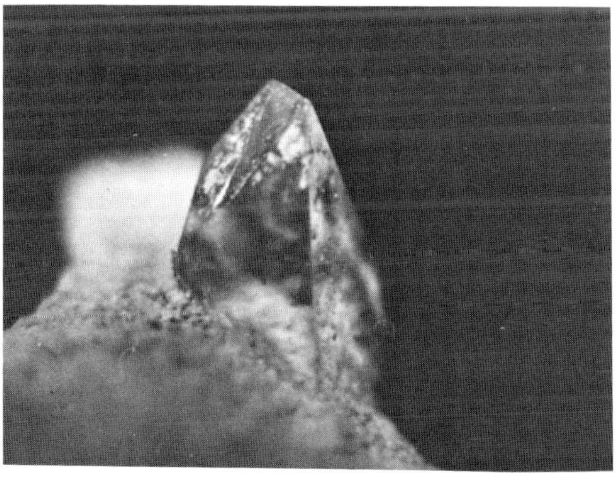
3

4

5

6

CHAPITRE I (Planches 1 à 9). *CHAPTER I (Plates 1 to 9)*

Un exemple de sédimentation évaporitique continentale actuelle. Le bassin tchadien.

An example of recent continental evaporitic sedimentation. The Chadian Basin (Africa).

PLANCHE 3 LES DEUX FACIES DE CRISTALLISATION DE LA MAGADIITE

Fig. 1 Lité → syngénétique (par évaporation directe d'une saumure affleurante).

Fig. 2 Nodulaire → diagénétique (par évaporation capillaire d'une nappe peu profonde, très concentrée).

(*Cliché* G. Maglione).

PLATE 3 BOTH FACIES OF MAGADIITE CRYSTALLIZATION

Fig. 1 Layered → syngenetic (by direct evaporation from on-exposed brine).

Fig. 2 Nodular → diagenetic (by capillary evaporation of a shallow, very concentrated nappe).

(*Photographer* G. Maglione).

2

CHAPITRE I (Planches 1 à 9). *CHAPTER I (Plates 1 to 9)*

Un exemple de sédimentation évaporitique continentale actuelle. Le bassin tchadien.

An example of recent continental evaporitic sedimentation. The Chadian Basin (Africa).

PLANCHE 4 MAGADIITE NODULAIRE RAPPELANT LE FACIES "CHICKEN-WIRE"

Dans le tiers inférieur de la photo remarquer les fines laminations de magadiite syngénétique.

(*Cliché* G. Maglione).

PLATE 4 NODULAR MAGADIITE WHICH RECALLS "CHICKEN-WIRE" FACIES

Note the fine laminations of syngenetic magadiite in the bottom third of the photo.

(*Photographer* G. Maglione).

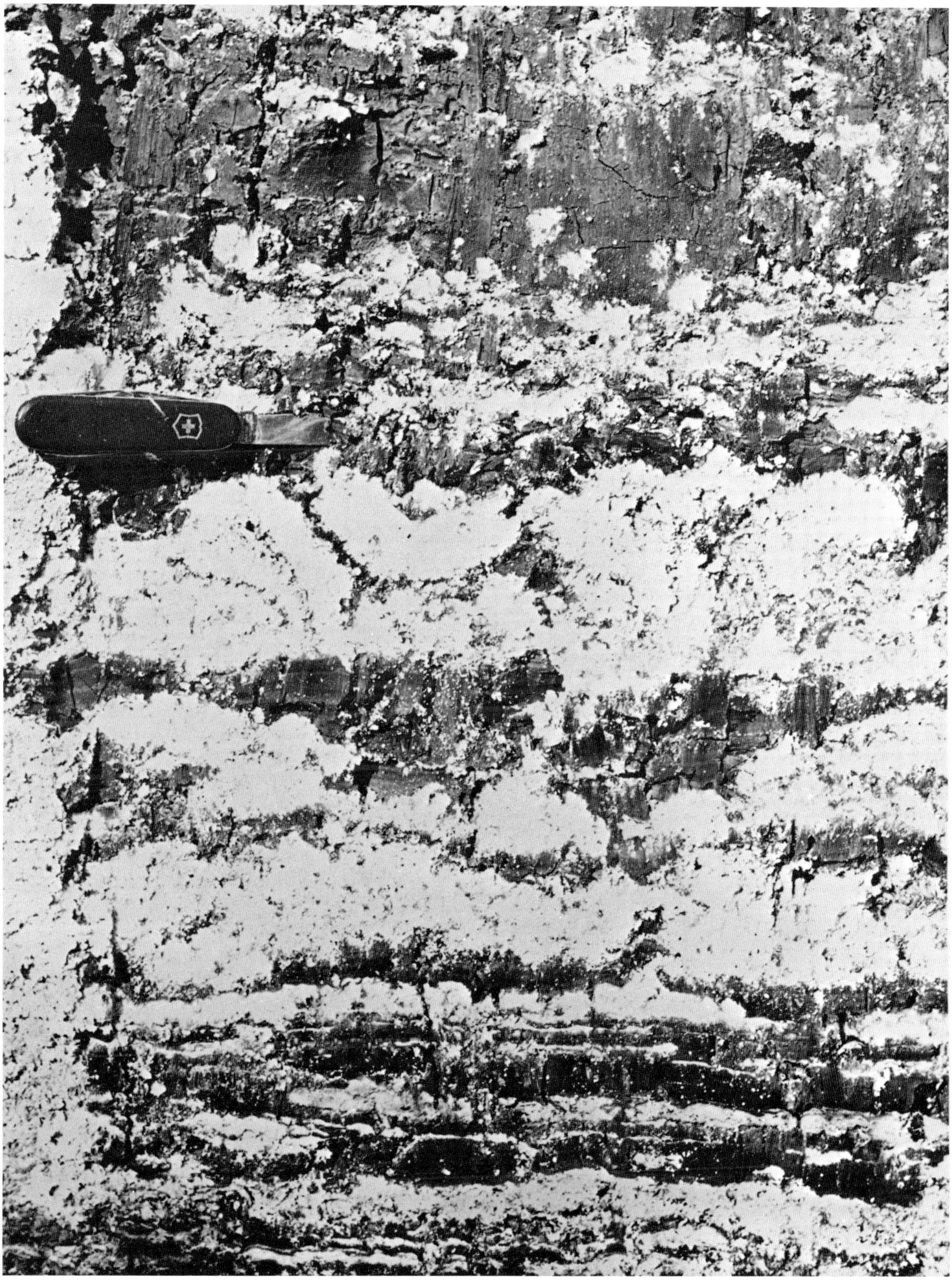

PLANCHE 5 MICROSCOPIE ELECTRONIQUE A BALAYAGE DE LA KANEMITE

Fig. 1 x 50.

Fig. 2 x 1 500.

Fig. 3 x 5 000.

(*Cliché Bureau Recherches Géologiques et Minières,* Orléans).

PLATE 5 SCANNING ELECTRON MICROSCOPY OF KANEMITE

Fig. 1 x 50.

Fig. 2 x 1 500.

Fig. 3 x 5 000.

(*Photography of Geological and Mining Research Bureau,* Orléans).

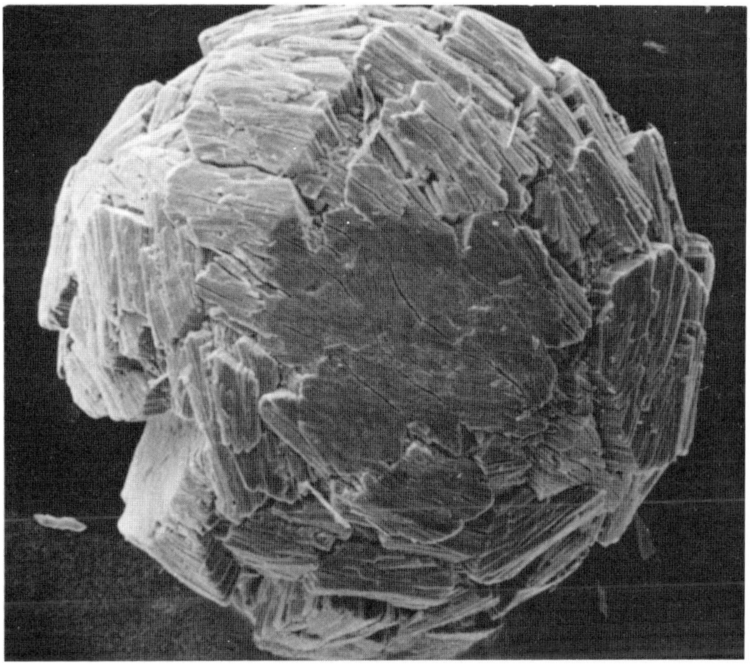

1

2

3

PLANCHE 6 MAGADIITE LITEE SYNGENETIQUE DU LAC MAGADI, KENYA (PLEISTOCENE SUPERIEUR).

(*Cliché* H.P. Eugster).

PLATE 6 LAYERED SYNGENETIC MAGADIITE OF LAKE MAGADI, KENYA (UPPER PLEISTOCENE).

(*Photographer* H.P. Eugster).

CHAPITRE I (Planches 1 à 9). *CHAPTER I (Plates 1 to 9)*

Un exemple de sédimentation évaporitique continentale actuelle. Le bassin tchadien.

An example of recent continental evaporitic sedimentation. The Chadian Basin (Africa).

PLANCHES 7 à 9

Dans les interdunes du Kanem, toutes ces figures sédimentaires sont représentées. Cependant à la différence du lac Magadi, elles affectent un matériau uniquement constitué de magadiite.

Ces figures ne sont donc pas nécessairement le résultat de la transformation de magadiite en chert, par l'intermédiaire ou non de kenyaite.

Dans le cas du Kanem, elles peuvent être dues à l'expulsion des saumures interstitielles lors de la lapidification de la magadiite plastique.

PLANCHE 7 EMPREINTES DE CRISTAUX A LA SURFACE DE DALLES DE CHERTS

Echantillon 298 : empreintes de cristaux de gaylussite maintenant occupés par un agrégat polycristallin de calcite noire (pseudogaylussite).
Echantillon en bas à gauche : empreintes de cristaux de trona, en aiguilles.
Lac Magadi, Kenya.

(*Cliché* H.P. Eugster).

PLATES 7 TO 9

All these sedimentary features are to be found in the interdunes of the Kanem. Unlike Lake Magadi, however, they only affect material made of magadiite.

These features do not, therefore, necessarily result from the transformation of magadiite into chert, with or without the influence of kenyaite.

In the case of the Kanem, they could be due to the expulsion of interstitial brines during lapidification of plastic magadiite.

PLATE 7 CRYSTAL PRINTS ON THE SURFACE OF CHERT SLABS

Sample 298 : gaylussite crystal prints now occupated by a policristalline aggregate of black calcite (pseudogaylussite).

Sample bottom left : traces of needle shaped trona crystals.
Lake Magadi, Kenya.

(*Photographer* H.P. Eugster).

298

PLANCHE 8 EMPREINTES DE FIGURES DE DESSICATION ET CHERTS TUBULAIRES

Fig. 1 Empreintes de figures de dessication à la surface d'une dalle de chert provenant de l'altération de magadiite ou de kenyaite.
Lac Magadi, Kenya.

Fig. 2 Cherts tubulaires, provenant vraisemblablement de dégagement gazeux dans un matériau encore plastique.
Lac Magadi, Kenya.

(*Clichés* H.P. Eugster)

PLATE 8 TRACES OF DESSICCATION MARKS AND TUBULAR CHERTS

Fig. 1 Traces of dessiccation marks on the surface of a chert slab, resulting from the change of magadiite or kenyaite.
Lake Magadi, Kenya.

Fig. 2 Tubular cherts probably resulting from gaseous escape in a still plastic material.
Lake Magadi, Kenya.

(*Photographer* : H.P. Eugster)

1

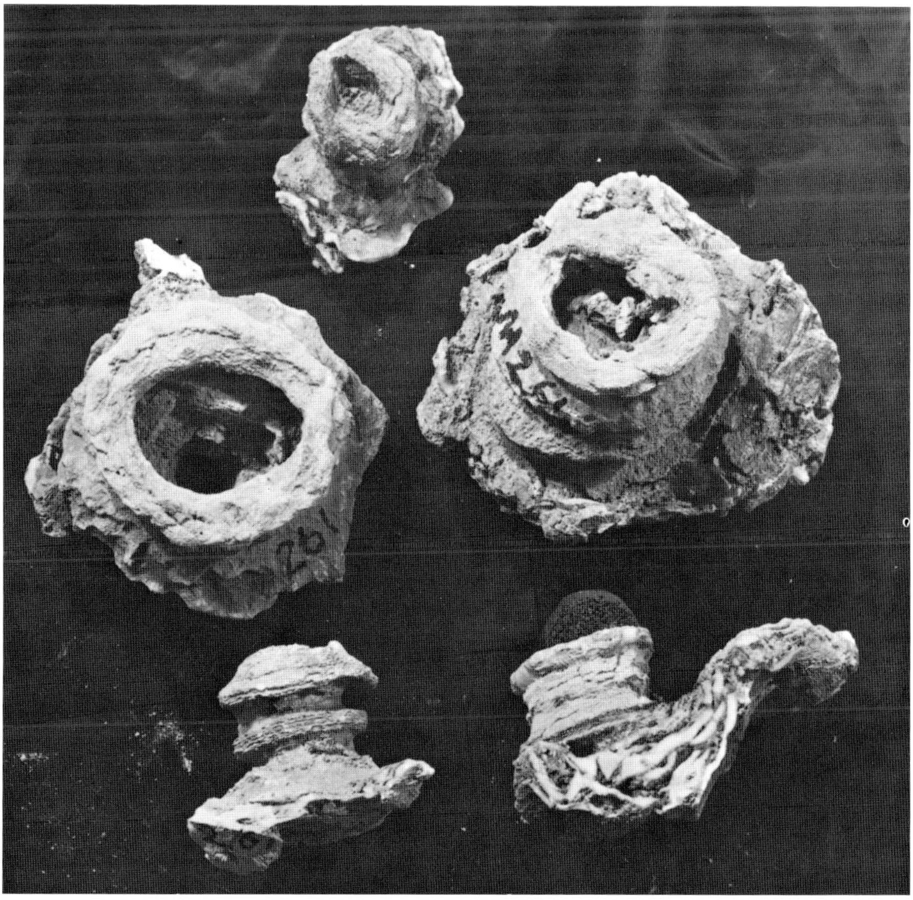

2

PLANCHE 9 SURFACE RETICULEE ET FIGURES DE DESSICCATION DANS DES DALLES DE CHERTS

Ces figures sont vraisemblablement dues à la contraction (au moins 25 %) qui accompagne la transformation de la magadiite en chert.
Lac Magadi, Kenya.

(*Cliché* H.P. Eugster)

PLATE 9 RETICULATE SURFACE AND DESSICATION MARKS IN CHERT SLABS

These marks are probably the result of the contraction (at least 25 %) which occurs when magadiite is transformed into chert.
Lac Magadi, Kenya.

(*Photographer* : H.P. Eugster)

CHAPITRE II (Planches 10 et 11). *CHAPTER II (Plates 10 and 11)*

La sebkha el Melah de Zarzis, bassin salin paralique subactuel (Tunisie).

Sebkha el Melah near Zarzis : a recent paralic salt basin (Tunisia).

PLANCHE 10 CONSTRUCTIONS ALGAIRES EN BORDURE DE LA SEBKHA EL MELAH

C'est probablement vers le début de la sédimentation évaporitique, dans les premiers stades de confinement du bassin, que les bordures de celui-ci furent colonisées par des algues bleues, responsables de ces constructions annulaires (on peut y rencontrer d'autres formes) qui ressemblent étrangement aux colonies stromatolithiques du Grand Lac Salé de l'Utah.

PLATE 10 ALGAL BUILDUP ON THE MARGIN OF SEBKHA EL MELAH

These ring-shaped algal buildings (one can find other types) were probably formed during the earliest stages of confinement. They surprisingly resemble the stromatolitic settlements in Great Salt Lake (Utah).

1

2

CHAPITRE II (Planches 10 et 11). *CHAPTER II (Plates 10 and 11)*

La sebkha el Melah de Zarzis, bassin salin paralique subactuel (Tunisie).

Sebkha el Melah near Zarzis : a recent paralic salt basin (Tunisia).

PLANCHE 11 PAVAGE DE CRISTAUX DE GYPSE POST-SEDIMENTAIRES, DEGAGE PAR DES ECOULEMENTS EN NAPPE, DANS LA ZONE D'ALIMENTATION DU BASSIN, AU DEBOUCHE DU CHENAL

Après émersion du seuil, l'alimentation du bassin en eau de mer se fait par inferoflux au sein des sédiments organo-détritiques du chenal.

Cette alimentation permet une concentration progressive des eaux qui y transitent, par évaporation capillaire. Ainsi une partie des sels peut se trouver piégée, les carbonates d'abord, le gypse ensuite et éventuellement la halite, mais celle-ci subit un lessivage par les eaux météoriques et est entraînée vers le centre du bassin. Si les carbonates restent discrets et le sel éphémère, le gypse, lui, s'individualise sous forme de grands cristaux automorphes qui poussent à l'intérieur du sédiment carbonaté, formant un véritable pavage superficiel, susceptible d'être dégagé par des écoulements en nappes à la suite de pluies torrentielles.

PLATE 11 PAVEMENT OF POST-SEDIMENTARY GYPSUM CRYSTALS, CLEARED OUT BY SHEET FLOODS, IN THE ZONE OF THE BASIN WHICH IS SEA WATER SUPPLIED (CHANNEL INLET).

After the sill emersion, the sea water supply into the basin is done by an underflow through the organo-detrital sediments of the channel.

This process allows the sea water travelling to the basin to gradually concentrate under effect of capillary evaporation. Thus, salts are partially trapped, carbonates first, then gypsum and eventually halite. But halite is washed away by rain waters towards the central parts of the basin. Thus, while halite is transitory and carbonates remain hidden within the sediment, gypsum accumulates and forms large automorphous crystals growing in the carbonate sediment. These crystals form a real pavement which can be cleared out by sheet floods after heavy rain.

1

2

CHAPITRE III (Planches 12 à 17). *CHAPTER III (Plates 12 to 17)*

Les faciès évaporitiques de sebkha.

Sebkha facies evaporites.

PLANCHE 12 SECTION DE CAROTTE DU PURBECKIEN INFERIEUR, WARLINGHAM, SURREY, ANGLETERRE

Cette planche illustre une partie d'un forage carotté dans la partie inférieure du Purbeckien inférieur.

La carotte montre une succession de bas en haut de dépôts de lagon, puis de tapis algaires renfermant des pseudomorphoses de gypse par la calcite ; au-dessus de ces tapis algaires, en faciès supratidal, se trouvent des séries de nodules d'anhydrite, dans un fond de dolomicrite. Le cycle est terminé par une surface d'érosion, au-dessus de laquelle existe un conglomérat de copeaux de boue carbonatée marquant la base des carbonates de lagon du prochain cycle.

(*Institut des Sciences Géologiques,* Londres, avec l'aimable autorisation du Directeur).

PLATE 12 CORE SECTION FROM THE LOWER PURBECK, WARLINGHAM, SURREY, ENGLAND

This plate illustrates part of a borehole core from the lower part of the Lower Purbeck.

The core shows an upward succession from lagoonal sediments into an algal mat which carries calcite pseudomorphs after gypsum ; above the algal mat, in the supratidal facies, courses of nodular anhydrite occur in a background of microdolomite. The cycle is terminated by an erosion surface, above which there is a carbonate mud-chip conglomerate which marks the base of the lagoonal carbonates of the next cycle.

(By courtesy of the Director, *The Institute of the Geological Sciences,* London).

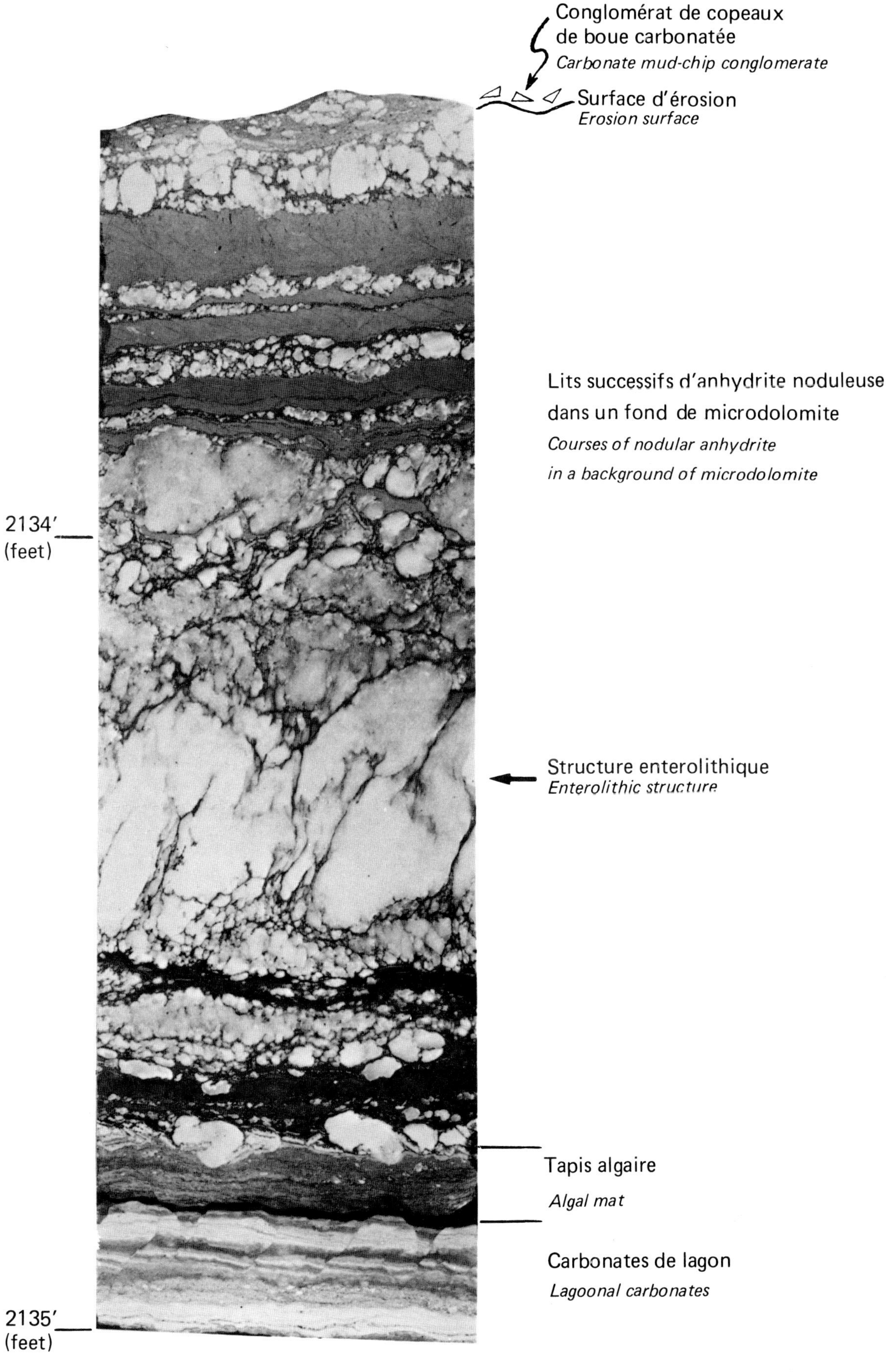
Conglomérat de copeaux de boue carbonatée
Carbonate mud-chip conglomerate
Surface d'érosion
Erosion surface
Lits successifs d'anhydrite noduleuse dans un fond de microdolomite
Courses of nodular anhydrite in a background of microdolomite
2134' (feet)
Structure enterolithique
Enterolithic structure
Tapis algaire
Algal mat
Carbonates de lagon
Lagoonal carbonates
2135' (feet)

CHAPITRE III (Planches 12 à 17). *CHAPTER III (Plates 12 to 17)*

Les faciès évaporitiques de sebkha.

Sebkha facies evaporites.

PLANCHE 13 STROMATOLITHES

Dans certains niveaux du Purbeckien inférieur du sud-est de l'Angleterre, forage de Warlingham, des stromatolithes se présentent comme témoins d'anciens faciès à anhydrite nodulaire. Les carbonates de ces stromatolithes sont à structure grumeleuse et les laminations sont souvent virtuellement absentes. Cependant, la forme générale de ces structures en section polie sur carottes ne laisse que peu de doute sur leur origine stromatolithique.

(*Institut des Sciences Géologiques,* Londres, avec l'aimable autorisation du Directeur).

PLATE 13 STROMATOLITES

In parts of the Lower Purbeck succession of south-east England, Warlingham borehole, stromatolites acted as host for emplacement of nodular anhydrite. The carbonate of the stromatolites has a clotted fabric and lamination is often virtually absent. However, the overall shapes of the structures as seen on polished surfaces of cores leaves little doubt that the masses of clotted carbonate were stromatolitic.

(By courtesy of the Director, *The Institute of the Geological Sciences,* London).

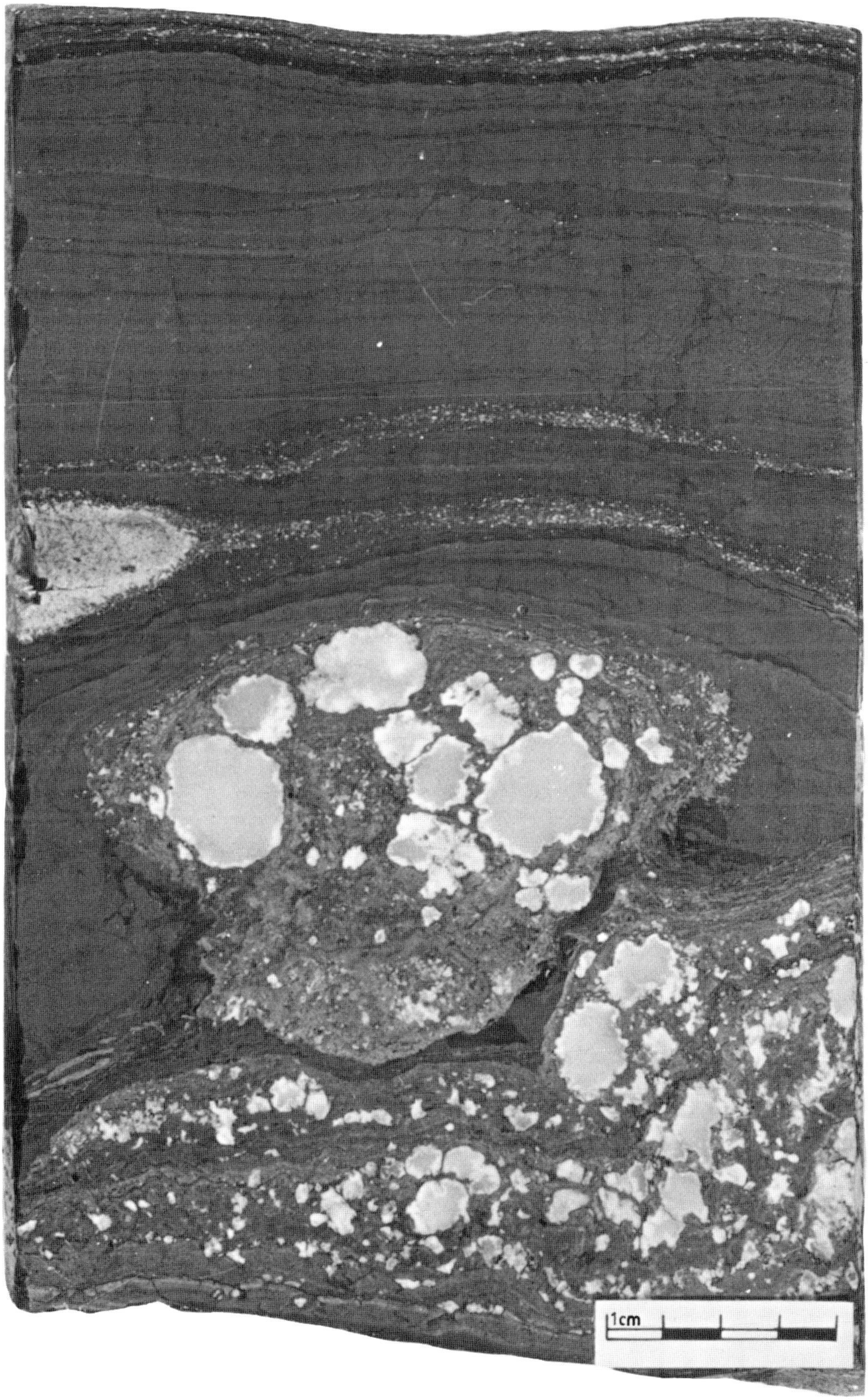
1cm

CHAPITRE III (Planches 12 à 17). *CHAPTER III (Plates 12 to 17)*

Les faciès évaporitiques de sebkha.

Sebkha facies evaporites.

PLANCHE 14 STROMATOLITHES

Des structures stromatolithiques actuelles comparables aux précédentes peuvent se trouver dans des lagunes côtières marines, par exemple à Laguna Mormona, Basse-Californie, Mexique, figures 1 et 2 (taille normale).

Ces stromatolithes présentent de petits amas et des lentilles de gypse. Peu ou pas de sédiment carbonaté est retenu par les stromatolithes, mais ils renferment d'abondants petits grains de carbonate de calcium qui se présentent comme ayant eu un développement diagénétique au sein des stromatolithes. Dans la partie inférieure des stromatolithes, où les grains carbonatés sont les plus développés, la structure est grumeleuse et la lamination originale a été oblitérée (fig. 2). On peut voir sur la photo que les lamines ne suivent pas exactement la forme des stromatolithes, mais les traversent plus ou moins horizontalement. Les stromatolithes se trouvent en groupes près des bords de la lagune et l'on ne sait pas s'ils correspondent à des structures en cours de développement ou si ce sont des structures reliques érodées de niveaux algaires plus anciens.

PLATE 14 STROMATOLITES

Comparable present day stromatolitic structures can be found in shallow saline coastal lagoons at Laguna Mormona, Baja California, Mexico, Figs. 1 and 2 (true size).

These stromatolites carry small pockets and lenses of gypsum. Little or no carbonate sediment is trapped by the stromatolites, but they carry abundant granules of calcium carbonate which appear to have grown diagenetically in them. In the lower parts of the stromatolites, where the carbonate granules are most abundantly developed, the structures have a clotted appearance and the original lamination has been obliterated, Fig. 2. It will be seen in the photograph that the laminae do not conform to the overall shape of the stromatolites, but run across them more or less horizontally. The stromatolites occur in clusters near the margin of the lagoon and it is uncertain whether they are growth structures or eroded relics of former spreads of algal mats.

1

2

CHAPITRE III (Planches 12 à 17). *CHAPTER III (Plates 12 to 17)*

Les faciès évaporitiques de sebkha.

Sebkha facies evaporites.

PLANCHE 15 STRUCTURES ENTEROLITHIQUES

Les structures entérolithiques sont un trait courant de beaucoup d'anciennes roches anhydritiques (exemple : Purbeckien inférieur, puits de Warlingham, sud-est de l'Angleterre).

Jusqu'à récemment il subsistait quelques incertitudes quant à leur origine. Cependant, les structures entérolithiques sont communément développées dans les anhydrites récentes des sebkhas d'Abu Dhabi (planche 16, fig. 2). Ces structures se forment quand des séries de nodules anhydritiques fusionnent pour former des couches et où le développement continu de ces nodules exige de l'espace. Les effets de pression latérale qui en résultent obligent les couches à devenir sinueuses. La formation de structures entérolithiques est sans aucun doute facilitée par le fort degré d'hydratation de ces nodules et l'habitus aplati des fins cristaux d'anhydrite qui glissent facilement entre eux.

(*Institut des Sciences Géologiques,* Londres, avec l'aimable autorisation du Directeur).

PLATE 15 ENTEROLITHIC STRUCTURES

Enterolithic structures are a common feature in many ancient anhydrite rocks (Example : Lower Purbeck, Warlingham borehole, southeast England) and until recently there was some uncertainty about their origin. However enterolithic structures are commonly developed in the recent anhydrite of the sebkhas of Abu Dhabi (plate 16, Fig. 2). The structures form where courses of anhydrite nodules have coalesced to form layers and continued growth of the nodules creates a demand for space. The lateral pressures which result cause the layers to become contorted. Production of enterolithic structures is no doubt facilitated by the high natural moisture content of the nodules, and the platy habit of the tiny anhydrite crystals which allows easy slip between them.

(By courtesy of the Director, *The Institute of Geological Sciences,* London).

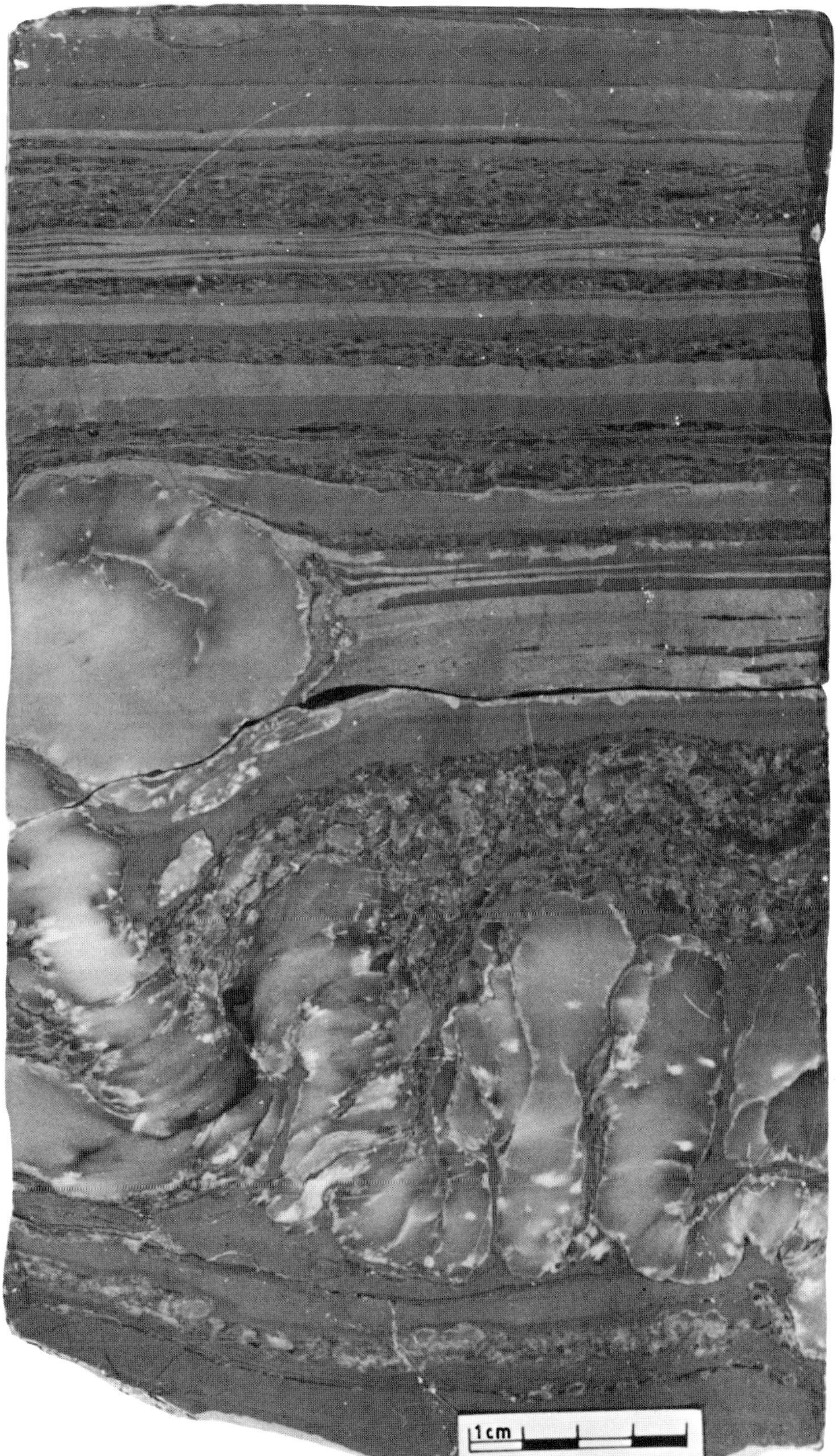
1cm

CHAPITRE III (Planches 12 à 17). *CHAPTER III (Plates 12 to 17)*

Les faciès évaporitiques de sebkha.

Sebkha facies evaporites.

PLANCHE 16 STRUCTURES ENTEROLITHIQUES

Les structures entérolithiques sont fréquentes dans les gypses secondaires observés à l'affleurement, par exemple dans les gypses issus d'hydratation tardive et superficielle d'anciennes anhydrites (fig. 1, Purbeckien inférieur, baie de Worbarrow, Dorset, Angleterre).

L'hydratation de l'anhydrite pour former le gypse aboutirait à un accroissement en volume de 63 %. Pour cette raison, les structures entérolithiques avaient été interprétées par le passé comme résultant de l'accroissement en volume engendré par l'anhydrite hydratée en gypse. Cependant, la comparaison de structures entérolithiques dans des gypses secondaires (fig. 1) avec celles d'une part observées dans des anhydrites (planche 15) et d'autre part avec celles qui se forment dans les sédiments récents (fig. 2, sebkha côtière, Abu Dhabi) ne laisse aucun doute sur l'origine primaire de ces structures de croissance.

PLATE 16 ENTEROLITHIC STRUCTURES

Enterolithic structures are common in secondary gypsum rocks seen at outcrop, i.e., in gypsum rocks which resulted from late, near-surface hydration of former anhydrite rocks : (Fig. 1, Lower Purbeck, Worbarrow Tout, Dorset Coast, England).

Hydration of anhydrite to form gypsum should result in an increase in volume of 63 %. For this reason it was suggested in the past that enterolithic structures formed in consequence of the increase in volume as anhydrite hydrated to gypsum. However comparison of enterolithic structures in secondary gypsum rocks (Fig. 1) with those in anhydrite rocks (Plate 15) and those which form in recent sediments (Fig. 2, coastal sebkha, Abu Dhabi) leaves no doubt that they are primary growth structures.

1

2

CHAPITRE III (Planches 12 à 17). *CHAPTER III (Plates 12 to 17)*

Les faciès évaporitiques de sebkha.

Sebkha facies evaporites.

PLANCHE 17 ANHYDRITE NODULAIRE ET SURFACE D'EROSION DANS DES DEPOTS RECENTS

(Voir texte de la légende page 108).

PLATE 17 RECENT NODULAR ANHYDRITE AND EROSION SURFACE IN RECENT ANHYDRITE

(See text of legend page 109)

1

2

CHAPITRE III (Planches 12 à 17). *CHAPTER III (Plates 12 to 17)*

Les faciès évaporitiques de sebkha.

Sebkha facies evaporites.

PLANCHE 17 ANHYDRITE NODULAIRE ET SURFACE D'EROSION DANS DES DEPOTS RECENTS

Fig. 1 Anhydrite nodulaire récente.

Dans beaucoup d'endroits des sebkhas du Sheikdom d'Abu Dhabi des séries de nodules d'anhydrite envahissent au moins la moitié de l'épaisseur des faciès supratidaux. Les nodules d'anhydrite croissent par déplacement du sédiment enveloppant et, en conséquence, ils soulèvent la surface de la sebkha.

L'anhydrite apparaît souvent en lits, mais ces lits sont en fait des assemblages de nodules coalescents.

Chaque nodule est un aggrégat de fins cristaux plats et croît non pas par développement des cristaux préexistants mais par naissance de nouveaux cristaux dans la matrice préexistante. Aussi, au fur et à mesure que de l'anhydrite est formée, cela demande-t-il de plus en plus d'espace, et des structures de croissance se développent ainsi que l'illustre le petit diapir de la planche. Sur le terrain, les nodules ont normalement un fort degré d'hydratation et sont mous, comme du fromage blanc, de telle sorte qu'ils s'étalent quand on les tranche en creusant. En conséquence, la structure nodulaire n'est pas visible sur les photographies de terrain.

(*Photographie,* Dr. G. Butler).

Fig. 2 Surface d'érosion de couches anhydritiques récentes.

Les sebkhas côtières d'Abu Dhabi s'étendent dans des parties basses, généralement à moins d'un mètre sous le niveau de haute mer et, en conséquence, sont largement inondées par la mer quand les vents côtiers coïncident avec les hautes marées de printemps.

Du sédiment est apporté et déposé sur la surface de sebkha pendant des périodes d'inondation calme et les vents y apportent des poussières et du sable, mais l'accrétion qui en résulte est un long et lent processus.

Cependant des vents très forts érodent périodiquement des étendues sèches et sableuses, tandis que les inondations violentes qui peuvent accompagner de grosses tempêtes peuvent enlever plusieurs centimètres de la surface de la sebkha en quelques heures. Aussi, la surélévation de la plaine de sebkha exprime-t-elle l'équilibre existant entre de longues périodes d'accrétion lente et des épisodes érosifs courts et violents.

Les preuves de ces événements sont nombreuses et l'une d'elles est spectaculairement illustrée par la troncature de couches anhydritiques sinueuses surmontées de nouveaux dépôts de sédiment (fig. 2, cliché Dr. G. Butler). Des surfaces d'érosion similaires ont été trouvées au toit des séries anhydritiques dans des carottes de sondages d'anhydrite noduleuse d'anciens faciès de sebkha (voir planche 12).

(*Photographie,* Dr. G. Butler).

PLATE 17 RECENT NODULAR ANHYDRITE AND EROSION SURFACE IN RECENT ANHYDRITE

Fig. 1 Recent nodular anhydrite.

In many parts of the coastal sebkhas of the Sheikdom of Abu Dhabi courses of nodular anhydrite make up at least half the thickness of the supratidal facies. The anhydrite nodules grow displacively in the sediments, and in consequence they must jack-up the surface of the sebkha.

The anhydrite often appears to be bedded, but it is not bedded in the true sense of sedimentary bedding. The layers are in fact courses of coalesced nodules.

Each nodule comprises an aggregate of tiny plate-like crystals and the nodules appear to grow, not by enlargement of the existing crystals, but by initiation of new crystals within the framework of earlier-formed ones. Thus as more and more anhydrite is emplaced it causes a demand for space, and growth structures develop, such as the small diapir illustrated in Plate 17. In the field the nodules usually have a high natural moisture content and are soft, rather like cream cheese, so that they smear when trenches are dug through them. In consequence the nodular structure does not show up in the field photographs.

(*Photograph* by Dr. G. Butler).

Fig. 2 Erosion surface in recent anhydrite.

The coastal sebkhas of Abu Dhabi are low lying, commonly less than a metre above normal high tide level, and in consequence they can be extensively flooded by the sea when on-shore winds coincide with high spring tides. Sediment is carried in and deposited on the sebkha surface during periods of gentle marine flooding and the winds carry in dust and sand, but the accretion which results is a slow long term process. However, very strong winds periodically deflate dry sandy areas, while the violent flooding which may accompany severe storms can erode several centimetres from the surface of the sebkha in a matter of hours. Thus the elevation of the sebkha plain expresses the balance between long periods of slow accretion and short sharp episodes of erosion. Evidence of these events is abundant and is spectacularly displayed where contorted layers of anhydrite are truncated and overlain by new layers of sediment (Fig. 2).

Similar erosion surfaces are found at the top of anhydrite units in borehole cores from ancient sebkha facies nodular anhydrite rocks (see plate 12).

(*Photograph* by Dr. G. Butler).

CHAPITRE IV (Planches 18 à 30). *CHAPTER IV (Plates 18 to 30)*

Successions carbonatées, anhydritiques et salifères dans le Dévonien moyen de forages du nord-est de l'Alberta (Canada occidental).

Carbonated, anhydritic and saliferous sequences in the Middle Devonian of wells in North-Eastern Alberta (West Canada).

PLANCHE 18 SONDAGE 10.6.101.6 W5 : DIAGRAPHIES CAROTTES ET LAMINES PLISSOTEES

Fig. 1 γ-ray, densité et sonic.

Fig. 2 Carotte 4. Muskeg.

Fig. 3 Echantillon de lamines plissotées rapportées à des phénomènes de *slumps*.

Au-dessus d'un corps carbonaté (Keg River) ici épais de 39 m (130 pieds) se sont déposées des alternances d'anhydrite et de carbonates, constituant la base de la "formation Muskeg", sur une épaisseur de 61 m (204 pieds). Ce régime d'alternances est surmonté de sel gemme (non visible sur la coupe ci-jointe). La carotte 4 illustrée sur les planches 18 et 19 se situe dans le régime des alternances. Seules les boîtes 9 à 4 (1), de bas en haut, sont illustrées sur ces deux planches (flèches extrêmes).

Tout à fait à la base de la coupe illustrée (boîte 9) s'observe sur 1,30 m environ une anhydrite massive, très pure, blanche, veinée de noir.

Au-dessus sur 1 m environ (sommet de la boîte 9 et la plus grande partie de la boîte 8), les carbonates se mêlent à l'anhydrite. On remarque des laminations parfois plurimillimétriques, composées, les unes de carbonate, les autres d'anhydrite, avec, suivant les zones, dominance de l'un ou de l'autre de ces composants. La figure de détail (fig. 3) montre un échantillon agrandi provenant de la base de la boîte 8 où une masse surtout carbonatée semble "flotter" dans un fond anhydritique. Les plissotements, fréquents dans les carbonates de toute cette zone, apparaissent sur cet échantillon imputables à des glissements (*slumps*) : le plissotement s'opère en effet de façon souple, sans cassure des lamines carbonatées voire anhydritiques. On notera le contraste entre la partie inférieure bien laminée, et la partie supérieure où carbonate et anhydrite sont mêlés intimement sans se séparer en lamines.

Les 30 derniers centimètres (boîte 8) font partie de la récurrence d'anhydrite massive illustrée page suivante (voir boîte 7).

PLATE 18 BOREHOLE 10.6.101.6 W5 : WELL LOGGINGS, CORES AND LAMINITES.

Fig. 1 γ ray, density and sonic log.

Fig. 2 Core 4. Muskeg.

Fig. 3 Sample with crumpled laminae interpreted as slumps.

Above a 39 m (130 ft) thick carbonated sequence (Keg River), there are alternating sediments of anhydrite and carbonates, which make up the base of the "Muskeg formation" 61 m (204 ft) thick. These alternations are surmounted by rock salt (not observable in the adjoining section). Core 4 shown on plates 18 and 19 is located within the trend of alternations. Only boxes 9 to 4 (1) from bottom to top, are illustrated on these 2 plates (outer arrows).

At the very bottom of the showned section (box 9), a very pure white massive anhydrite with black veining can be seen over about 1,30 m. Above, over about 1 m at the top of box 9 and most of box 8, carbonates are mixed with the anhydrite. Plurimillimetric laminations can be distinguished, some of carbonate others of anhydrite, with predominance of the one or the other constituent depending on the zones. The detailed view, (3) is of a magnified sample from the base of box 8 where a principally carbonated mass seems to "float" in an anhydritic background. The crumplings – frequently seen in the carbonates in the whole of this zone – seem in this sample attributable to slumps : the crumpling in fact is effected with a certain suppleness without fractures in carbonate or even anhydritic laminae. Note the contrast with the lower well-laminated portion and the upper part where carbonate and anhydrite are closely mixed without being separated into laminae.

The last 30 cm (box 8) are part of the recurrence of massive anhydrite seen on the next page (see box 7).

---

(1) Chaque boîte, haute de 90 cm (3 pieds), comprend 2 tronçons sciés (*slabs*) celui de droite étant originellement superposé à celui de gauche, dans l'ordre de la succession stratigraphique ; sur les clichés où se trouvent regroupés deux boîtes, l'une à droite avec le numéro le plus faible (8) est originellement susjacente à celle de gauche correspondant au numéro le plus fort (numérotation des boîtes *per descensum*).

(1) Each box, 90 cm (3 feet) high includes two slabs, the one on the right originally having been superimposed on the one of the left in stratigraphic order of succession.

Figures showing the two boxes : the one on the right with the lowest number (ex. 8) originally overlays the left hand one which has a higher number (boxes are numbered "descending order").

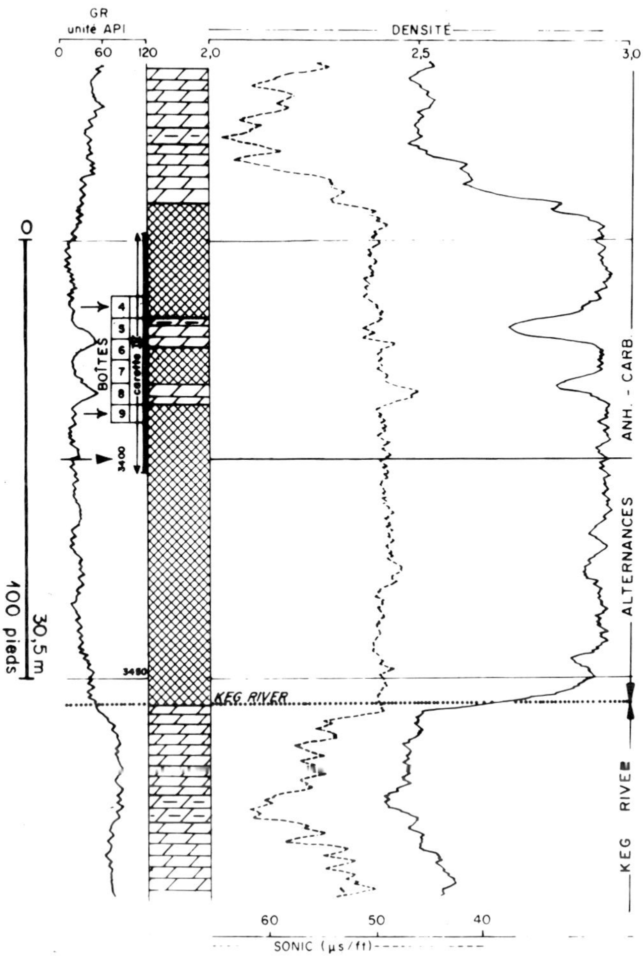
GR
unité API
0 60 120
DENSITÉ
2,0 2,5 3,0
BOÎTES
4 5 6 7 8 9
carotte
3400
3480
KEG RIVER
ANH. - CARB.
ALTERNANCES
KEG RIVER
60 50 40
SONIC (μs/ft)
0
30,5 m
100 pieds

1

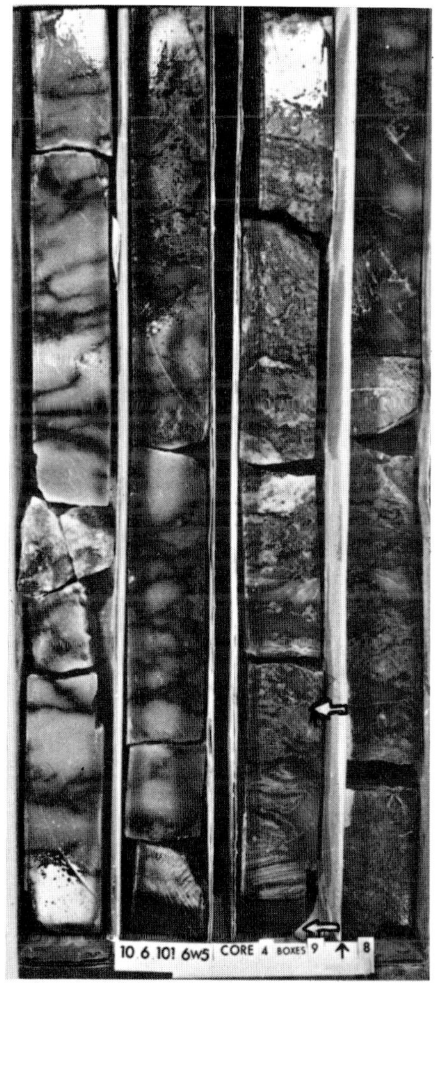
10 6 101 6W5 CORE 4 BOXES 9 8

2

cm
10 6 101 6W5
IV 8

3

CHAPITRE IV (Planches 18 à 30). *CHAPTER IV (Plates 18 to 30)*

Successions carbonatées, anhydritiques et salifères dans le Dévonien moyen de forages du nord-est de l'Alberta (Canada occidental).

Carbonated, anhydritic and saliferous sequences in the Middle Devonian of wells in North-Eastern Alberta (West Canada).

PLANCHE 19 SONDAGE 10.6.101.6 W5 (suite et fin) : SECTIONS DE CAROTTES ET LAMINITES

(Voir texte de la légende page 114)

PLATE 19 BOREHOLE 10.6.101.6 W5 (continued to end) : CORE SECTIONS AND LAMINITES

(See text of legend page 115).

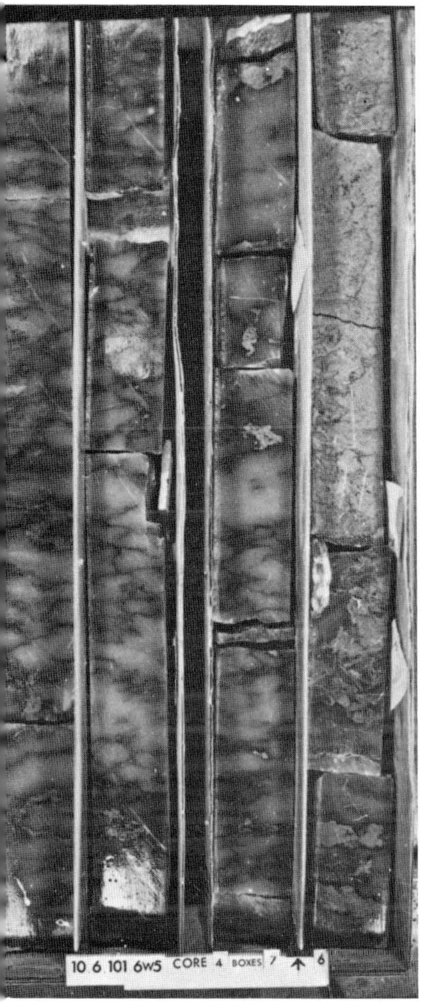
10 6 101 6w5 CORE 4 BOXES 7 ↑ 6

1

10.6.101.6w5 → IV 5
CENTIMETRES

a

2

10.6.101.6w5 → IV 5
CENTIMETRES

b

3

← b

ⓐ
10 6 101 6w5 CORE 4 BOXES 5 ↑ 4

4

PLANCHE 19 SONDAGE 10.6.101.6 W5 (suite et fin) : SECTIONS DE CAROTTES ET LAMINITES

Fig. 1 et 4 Sections de carottes de Muskeg.

Fig. 2 Passée de carbonates à laminations plan parallèles et anhydrite diagénétique (p.p.).

Fig. 3 Laminites dolomitiques inclinées.

Sur 3,40 m environ (sommet de la boîte 8, planche 18, boîte 7, boîte 6 jusqu'à 50 cm du sommet) de l'anhydrite massive où une structure "mosaïque" semble parfois résulter de phénomènes entérolithiques, mêle des zones très blanches et des zones bleu sombre à noirâtres. On note la présence de zones laminées et contournées, centimétriques, de carbonates. Les 20 cm sommitaux de cette anhydrite massive se distinguent par la présence de bancs carbonatés de quelques centimètres à contour sinueux. Cette intercalation d'anhydrite massive s'inscrit remarquablement sur les enregistrements gamma ray, sonic et densité (planche 18).

50 cm environ (sommet de la boîte 6 et extrême base de la 5) dolomie encore anhydritique, d'aspect pseudo-bréchique, qui s'interpose, dans une succession logique, entre l'anhydrite sous-jacente. Cette "bréchification" suggère des cassures, postérieures à la lithification, ayant affecté un matériel analogue aux laminites sus-jacentes.

1,20 m environ (soit la plus grande partie de la boîte 5) de dolomies à fines laminations plan-parallèles. On notera qu'à la succession ci-dessous décrite (anhydrite massive, anhydrite à corps carbonatés, carbonates pseudobréchiques) fait suite logiquement une succession comprenant laminites dolomitiques inclinées (sur environ 70 cm ; voir figure de détail 3) puis en concordance laminites dolomitiques horizontales ; cette succession générale est tout à fait habituelle et met en jeu des phénomènes évolutifs sédimentologiques et diagénétiques. Par ailleurs on notera qu'une certaine ondulation affecte le corps de laminites "calmes".

2,40 m (épaisseur minimale : fin de la boîte 5 et boîte 4, dernière à être présentée ici) d'anhydrite massive, à structure en mosaïque, bleu sombre, pratiquement homogène à l'exception de fines lamines d'épaisseur millimétrique, disposées en chevelu et constituées de carbonate clair. Au sommet des 20 premiers centimètres de cette anhydrite massive on observe une récurrence de carbonates finement feuilletés (boîte 5 et détail figure 2). Cette passée épaisse de 3,5 cm composée de laminations millimétriques plan parallèles, est séparée de l'anhydrite sous-jacente par contact irrégulier. A la partie supérieure, le contact avec l'anhydrite sus-jacente est franc et plus parallèle aux laminations. L'analogie est frappante entre cette récurrence et les laminites sous-jacentes. On peut émettre l'hypothèse que l'anhydrite massive intercalée correspond, au moins en partie, à un développement diagénétique.

PLATE 19 BOREHOLE 10.6.101.6 W5 (continued to end) : CORE SECTIONS AND LAMINITES

Figs. 1 and 4 Muskeg slabed cores.

Fig. 2 Interbedding of parallel lying carbonate laminations and diagenetic anhydrite.

Fig. 3 Dolomitic inclined laminae.

Throughout about 3.40 m (top of box 8, plate 18 ; box 7, box 6, up to 50 cm from the top), massive anhydrite where a mosaïc structure seems sometimes to emanate from enterolitic phenomena, mixes very white zones and dark blue to black zones. We observe the presence of centimetric laminated and distorted carbonated zones. The 20 top centimeters of this massive anhydrite can be distinguished by the presence of carbonate beds of several centimeters with sinuous edges. This interbedding of massive anhydrite is seen particularly on the gamma ray, sonic and density log recordings (Plate 18).

About 50 cm (at the top of box 6 and very bottom of box 5) of still anhydritic dolomite of pseudo-brecciated appearance, interposing in a logical succession between the underlying anhydrite and overlying finely laminated carbonates. This "breccification" suggests breaks occurring after lithification, after affecting material analogous to the overlying laminites.

About 1.20 m (most of box 5) of parallel-lying, finely laminated dolomites. After the succession described below (massive anhydrite, anhydrite with carbonate layers pseudo-brecciated carbonates) we note the immediate follow-on of a succession of inclined dolomitic laminites (over about 70 cm ; see detailed view (Fig. 3), then, in conformity horizontal dolomitic laminites ; this general succession is completely normal and involves sedimentological and diagenetic evolutionary features. Furthermore, we notice one ondulation affecting the "calm" laminate layers.

2.40 m (minimal thickness, end of boxes 5 and 4, last ones presented here) of massive anhydrite with mosaic structure, dark blue and practically homogenous apart from fine, millimetric anastomosed laminae and made up of light-coloured carbonate. At the top of the first 20 cm of that massive anhydrite, finely laminated carbonates recur (box 5 and detailed view (Fig. 2)). This 3.5 cm thick sequence composed of millimetric, parallel-lying laminations, is separated from the underlying anhydrite by irregular limits. In the upper part, the boundary with the overlying anhydrite is clear-cut and more parallel to the laminations. The similarity is striking between this recurrence and the underlying laminites. The hypothesis may be put forward that interbedded massive anhydrite corresponds, at least in part, to a diagenetic development.

CHAPITRE IV (Planches 18 à 30). *CHAPTER IV (Plates 18 to 30)*

Successions carbonatées, anhydritiques et salifères dans le Dévonien moyen de forages du nord-est de l'Alberta (Canada occidental).

Carbonated, anhydritic and saliferous sequences in the Middle Devonian of wells in North-Eastern Alberta (West Canada).

PLANCHE 20 SONDAGE 6.30.86.13 W4 : SECTIONS DE CAROTTES, DIAGRAPHIES ET FACIES A FAUNE BENTHIQUE

(Voir texte de la légende page 118).

PLATE 20 BOREHOLE 6.30.86.13 W4 : CORE SECTIONS, WELL LOGGINGS AND FACIES WITH BENTHIC FAUNA

(See text of legend page 119).

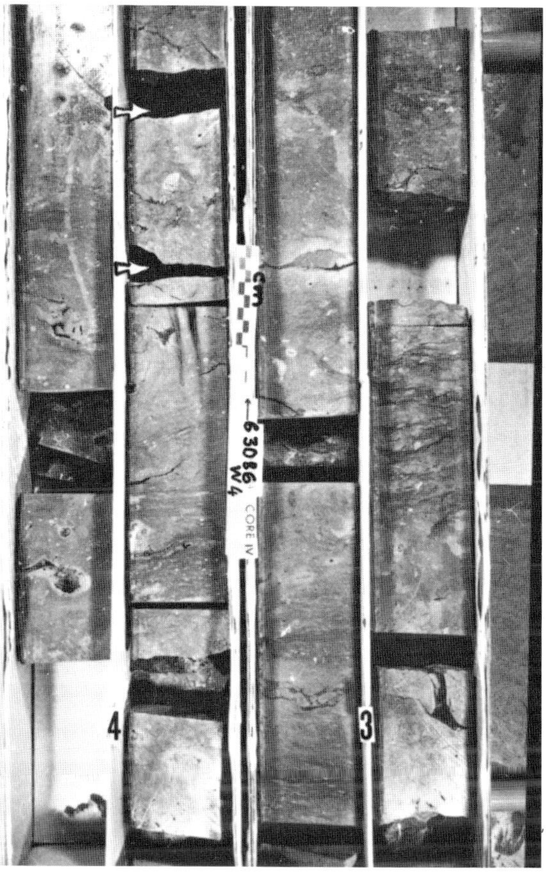
cm
6 30 86 W4
CORE IV
4
3

1

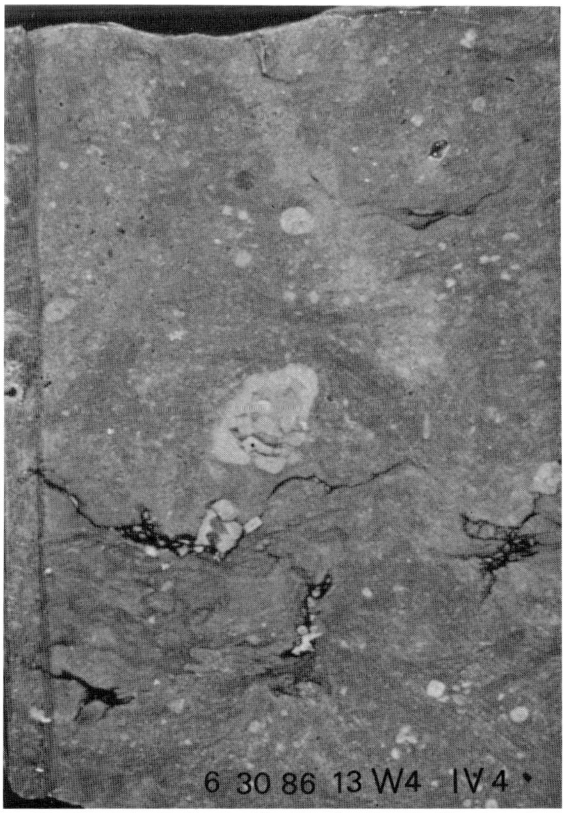
6 30 86 13 W4 IV 4

2

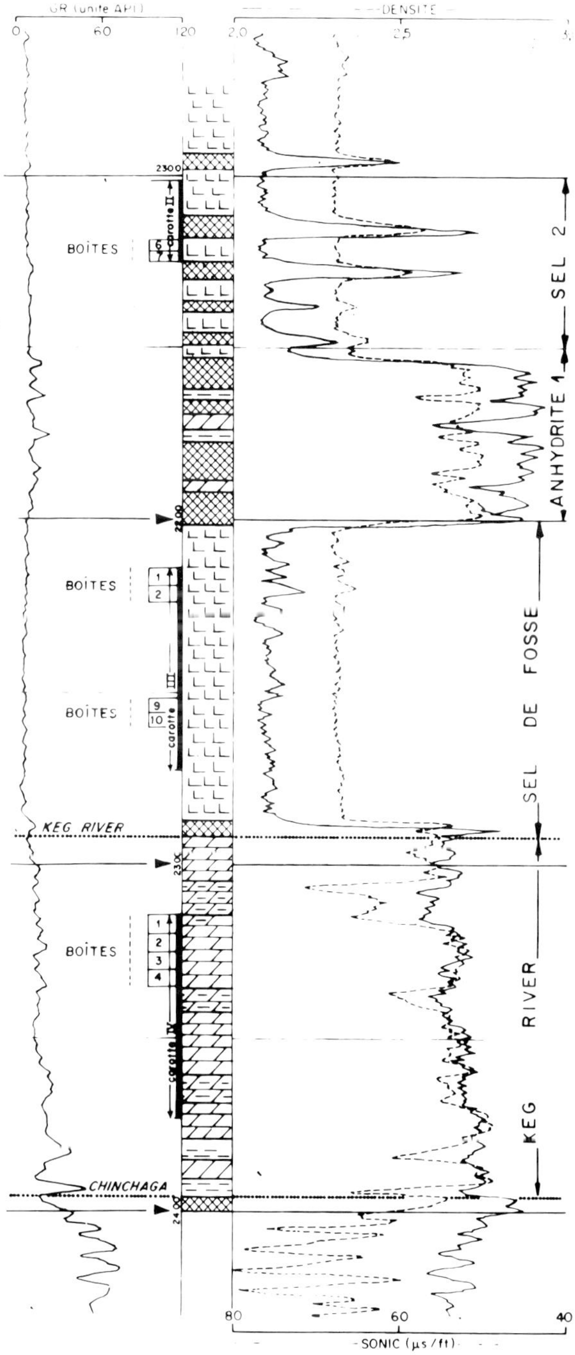
GR (unité API)
0
60
120
DENSITE
2,0
2,5
3
BOÎTES
carotte II
BOÎTES
BOÎTES
carotte III
KEG RIVER
BOÎTES
carotte IV
CHINCHAGA
2300
2400
SEL 2
ANHYDRITE 1
SEL DE FOSSE
KEG RIVER
80
60
40
SONIC (µs/ft)

3

CHAPITRE IV (Planches 18 à 30). *CHAPTER IV (Plates 18 to 30)*

Successions carbonatées, anhydritiques et salifères dans le Dévonien moyen de forages du nord-est de l'Alberta (Canada occidental).

Carbonated, anhydritic and saliferous sequences in the Middle Devonian of wells in North-Eastern Alberta (West Canada).

PLANCHE 20 SONDAGE 6.30.86.13 W4 : SECTIONS DE CAROTTES, DIAGRAPHIES ET FACIES A FAUNE BENTHIQUE

Fig. 1 Sections de carottes Keg River.

Fig. 2 Faciès à faune benthique bioturbé.

Fig. 3 γ ray, densité, sonic du Keg River et Muskeg.

Ce forage est situé en position intermédiaire entre deux domaines différents :

A l'ouest, un domaine de plate-forme à l'époque Keg River, a vu la sédimentation de calcaires relativement épais, peu stratifiés, riches en faunes benthiques, à faciès "oxygéné". Ces calcaires de "type plate-forme" sont surmontés par des évaporites, d'extension très générale ("sel de nappe").

Au contraire, à l'est, se développe une "fosse" où les épaisseurs **globales** sont un peu plus fortes. Les calcaires du Keg River s'amincissent dans cette "fosse" jusqu'à ne plus faire qu'une dizaine de mètres en son centre et tendent à y devenir finement laminés, plus riches en matière organique mais dépourvus de débris benthiques ; bref, présentant un faciès réducteur. Au-dessus de ce Keg River aminci, la fosse se trouve remplie par un sel très pur (voir planche 22). A la fin du dépôt de ce sel très pur, il y a eu nivellement avec le domaine plate-forme et du sel de nappe s'est déposé uniformément (voir planche 23).

Le forage, ici illustré fig. 3, appartient déjà plutôt au domaine de fosse : les calcaires Keg River ne font qu'une trentaine de mètres (104 pieds) et, au-dessus, le sel de fosse est épais de 27 m (90 pieds). Une excellente définition des ensembles lithologiques successifs est fournie par les gamma ray, sonic et densité. Le diamétreur (non reproduit sur la planche 20) atteste la très bonne qualité du trou, sans dissolution importante même dans les intervalles salifères épais.

Le début de la zone illustrée (base de la boîte 4 de la carotte 4) se situe à 19,5 m (65 pieds) de la base du Keg River (contact avec la formation évaporitique sous-jacente, dite du Chinchaga). On se trouve là dans les carbonates de faciès plate-forme et ceux-ci se prolongent sur un peu plus de 6,50 m, couvrant les boîtes 4, 3 (planche 20), 2 et une partie de 1 (planche 21). Il s'agit de calcaires plus ou moins dolomitisés, de couleur beige à roux. La stratification est pratiquement inexistante. En effet la structure est le plus souvent noduleuse ou pseudobréchique, probablement sous l'effet de bioturbation, parfois évoquée par un certain fluage de la boue carbonatée. La disposition tout à fait anarchique, à l'exclusion de toute disposition en lits, des débris d'organismes s'ajoute pour souligner cette absence de stratification. Ces organismes consistent principalement, pour ceux qui sont visibles à l'œil nu, en débris de tiges d'encrines d'assez grosse taille à la base, (en particulier boîte 4, planche 20, photo d'ensemble 1 et figure de détail 2) et devenant plus petits au-dessus (voir sur planche 21, boîte de 2 photos d'ensemble (fig. 1) et de détail (fig. 2) et première moitié de la boîte 1). Il s'agit donc d'un faciès à faune benthique, brassé, oxygéné. Toutefois, on remarquera des passées en filets, subparallèles à la stratification, pouvant atteindre plusieurs centimètres de longueur, plus ou moins continues et régulières (la bioturbation peut être responsable des interruptions). Ces passées semblent plus riches en matière organique et évoquent déjà des tentatives d'installation d'un milieu plus réducteur alors que les trajets de bioturbation, plus clairs que la matrice, paraissent plus oxydés que celle-ci.

PLATE 20 BOREHOLE 6.30.86.13 W4 : CORE SECTIONS, WELL LOGGINGS AND FACIES WITH BENTHIC FAUNA

Fig. 1 Keg River slabed cores.

Fig. 2 Benthic fauna bioturbated facies.

Fig. 3 Keg River and Muskeg, γ ray, density and sonic logs.

This borehole lies between two different areas :

To the west, the platform environment in the Keg River period witnessed deposition of quite thick, little bedded limestones, rich in benthic fauna and with "oxygenised" facies. These "platform type" limestones are overlain by widespread evaporites ("sheet salt").

On the contrary, to the east, a "trough" is developing where overall thicknesses are greater. The Keg River limestones in this "trough" are thinning to no more than about ten meters in the center, becoming finely laminated, and richer in organic matter, but without benthic remains ; in other words, a reducing facies. Above this thinned Keg River, the "trough" becomes filled with very pure salt (see Plate 22). At the end of this deposit of pure salt there is levelling with the platform area, and sheet salt is deposited uniformly (see Plate 23).

The borehole illustrated here (Fig. 3) is part of the "deep" environment : the Keg River limestones cover only about 30 m (104 feet) and the "deep" salt above is 27 m (90 feet) thick. A clear definition of successive lithological series is given by gamma ray, sonic and density logs. The inner caliper (not reproduced on the adjoining plate) indicates the good quality of the hole, without much dissolution even in the thick saliferous intervals.

The start of the illustrated zone (bottom of box 4, core 4) is at 19.5 m (65 feet) from the bottom of Keg River (limit of the underlying evaporitic formation called Chinchaga). This is within the layer of carbonates with "platform" facies, which extend a little more than 6.50 m, boxes 4, 3 (Plate 20), 2 and some of box 1 (Plate 21). Here the limestones are more or less dolomitised and beige to red in colour. There is hardly any bedding. The structure is usually nodular or pseudo-brecciated, probably by bioturbation effect which is sometimes evoked by the flow of lime mud. The completely unordered arrangement – apart from in beds – of remains of organisms further underlines the lack of bedding. These organisms seen to the naked eye consist essentially of *Encrinus* remains which are quite large at the base (especially box 4, Plate 20, photo of the whole, Fig. 1 and detailed view, Fig. 2), becoming smaller higher up (see Plate 21, box 2, figures of the whole, (Fig. 1) and of the detailed view, (Fig. 2) and the first half of bow 1). It is therefore a winnowed and oxygenated facies with benthic fauna. Strata can be seen subparallel to the bedding, up to several centimeters long, more or less continuous and regular (bioturbation may cause interruptions). These zones seem to be richer in organic material and already suggest the possible establishment of a more reducing environment whereas bioturbation traces, which are lighter in colour than the matrix seem to be more highly oxydised.

CHAPITRE IV (Planches 18 à 30). *CHAPTER IV (Plates 18 to 30)*

Successions carbonatées, anhydritiques et salifères dans le Dévonien moyen de forages du nord-est de l'Alberta (Canada occidental).

Carbonated, anhydritic and saliferous sequences in the Middle Devonian of wells in North-Eastern Alberta (West Canada).

PLANCHE 21 SONDAGE 6.30.86.13 W4 (suite) : SECTIONS DE CAROTTES, CARBONATES LAMINES ET FACIES BIOTURBE.

Fig. 1 Sections de carottes du Keg River.

Fig. 2 Carbonates laminés plan-parallèles.

Fig. 3 Faciès carbonaté bioturbé.

Aux faciès carbonatés de plate-forme (boîte 2 et partie de la boîte 1, figure de détail 3), déjà décrits avec la planche précédente, font suite les carbonates laminés. Ceux-ci n'existent ici que sur une cinquantaine de centimètres. D'apparence typiquement pré-évaporitique, d'après d'autres observations, ils pourraient se poursuivre jusqu'au sel sus-jacent, soit sur une dizaine de mètres, malheureusement non carottés. On notera que ces faciès laminés ont été précédés dans les calcaires "plate-forme" sous-jacents par une diminution de la taille des organismes, ce qui paraît en accord avec une évolution cohérente.

Ces couches sont caractérisées par l'existence d'une très fine lamination (parfois infra-millimétrique), très régulière et plan-parallèle. Ce phénomène est particulièrement remarquable à la partie sommitale de la carotte (sommet de la boîte 1 et figure de détail 2), c'est-à-dire lorsque l'on s'éloigne verticalement des faciès "plate-forme" sous-jacents. Cette lamination consiste en alternances de lits beige clair (carbonatés) et brun à noir (plus riches en matière organique et/ou en argile). Elle est similaire à de nombreux cas de faciès réducteurs évoqués par ailleurs pour leur richesse en nannoplancton (Busson et Noël, 1972). Ces couches finement laminées correspondent, on le sait, à une sédimentation conditionnée par la superposition de deux corps d'eau, soumis éventuellement à des variations individuelles. On notera enfin la présence de lits plus épais (centimétriques), d'apparence plus homogène, où pourrait se manifester une discrète bioturbation (figure de détail 2). De tels lits ont été interprétés par Davis et Ludlam (1971, communication orale, et 1973) comme des dépôts de tempête, par comparaison avec les dépôts actuels de Green Lake (New York), d'aspect général très similaire.

PLATE 21 BOREHOLE 6.30.86.13 W4 (continued) : CORE SECTIONS, LAMINATED CARBONATES AND BIOTURBATED FACIES.

Fig. 1 Keg River slabed cores.

Fig. 2 Laminated plan parallel carbonates.

Fig. 3 Bioturbated carbonate facies.

Laminated carbonates follow the carbonated platform facies (box 2 and part of box 1, detailed view 3) already described in the previous plate. These laminated carbonates only cover about fifty centimeters and are of typical pre-evaporitic appearance according to other observations. It would be possible for them to continue up to the overlying salt, that is over another about ten meters, unfortunately not cored. Just before these laminated facies in underlying platform limestones, the organisms are seen to decrease in size, which supports the idea of a coherent evolution.

These layers are characterized by a very fine (sometimes infra-millimetric) regular and plan parallel lamination, which is particularly noticeable in the topmost part of the core (top of box 1 and detailed view 2), when away from the underlying "platform" facies. This lamination is made up of alternations between light beige (carbonated) beds, and brown to black ones (richer in organic matter and/or clays (?)). It resembles numerous examples of reducing facies noted elsewhere for their abundance of nannoplankton (Busson and Noël, 1972). These finely laminated layers correspond to sedimentation conditioned by the superposition of two water bodies, which eventually undergo individual variations. Lastly, thicker (centimetric) beds may be observed, which seem more homogeneous and may possibly show slight bioturbation (detailed view 2). Such layers were interpreted by Davies and Ludlam (1971, oral communication, and 1973) as storm deposits, in comparison with actual Green Lake deposits (New York), which generally appear to be very similar.

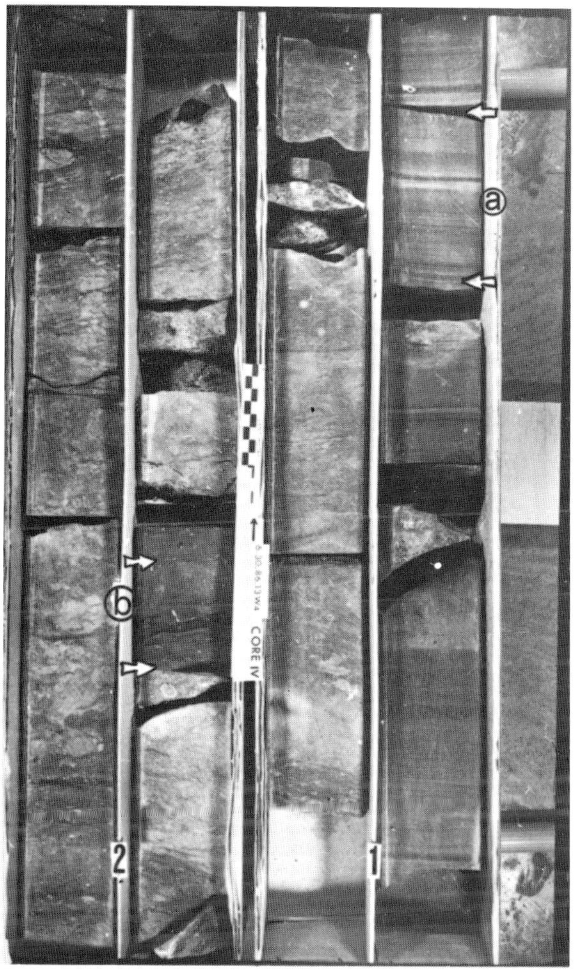

1

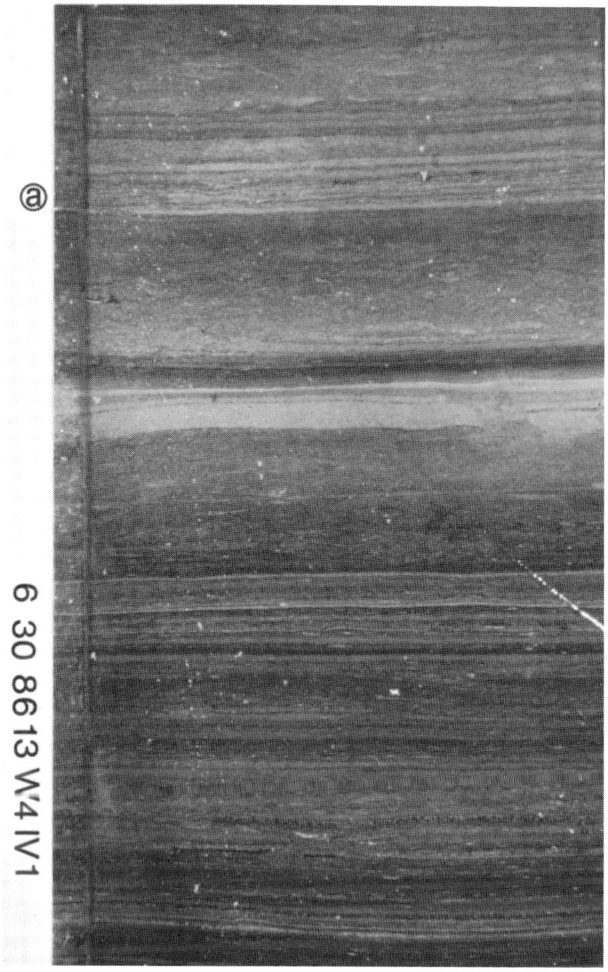

2

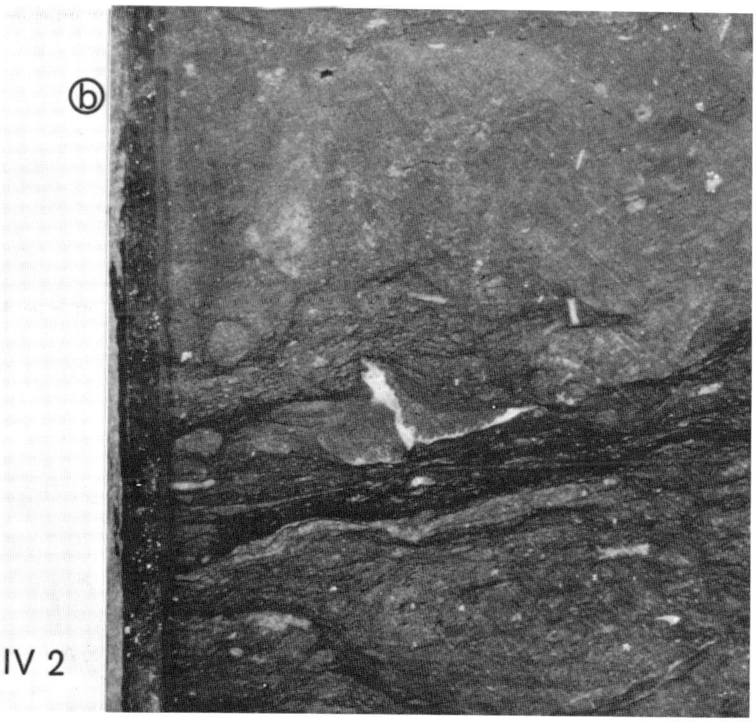

3

CHAPITRE IV (Planches 18 à 30). *CHAPTER IV (Plates 18 to 30)*

Successions carbonatées, anhydritiques et salifères dans le Dévonien moyen de forages du nord-est de l'Alberta (Canada occidental).

Carbonated, anhydritic and saliferous sequences in the Middle Devonian of wells in North-Eastern Alberta (West Canada).

PLANCHE 22 SONDAGE 6.30.86. 13 W4 (suite) : SECTIONS DE CAROTTES, CONTACT SEL/CARBONATES

(Voir texte de la légende page 124).

PLATE 22 BOREHOLE 6.30.86.13 W4 (continued): CORE SECTIONS, SALT/CARBONATES CONTACT

(See text of legend page 125).

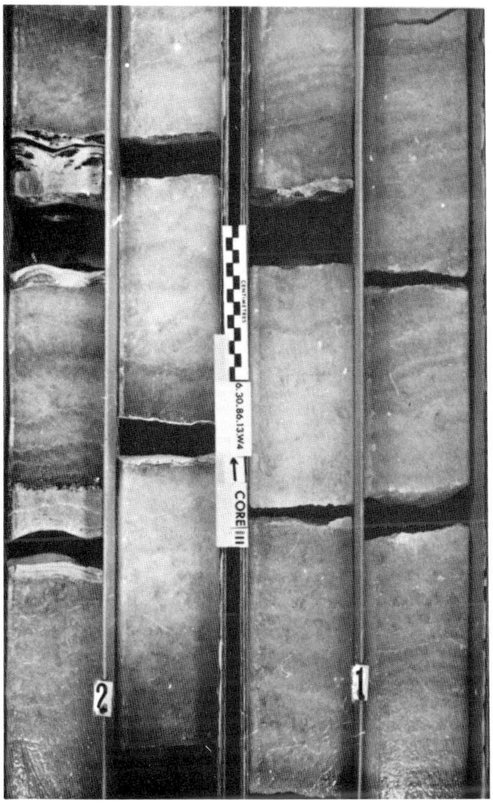

1

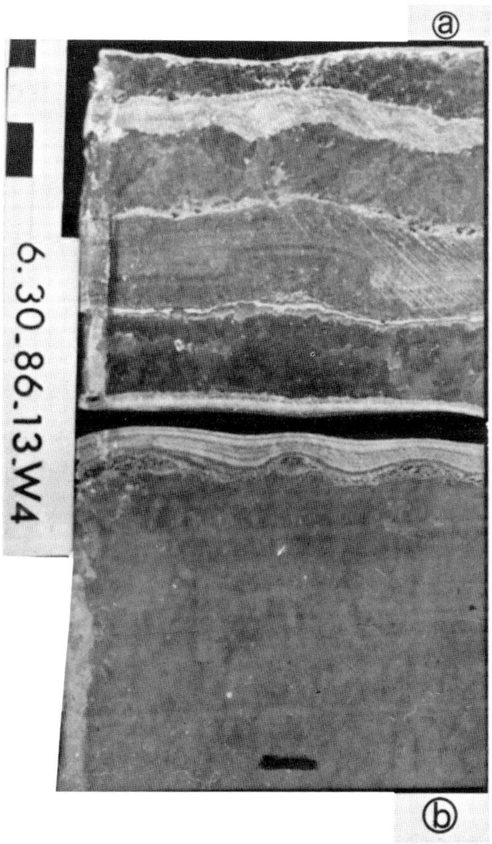

3

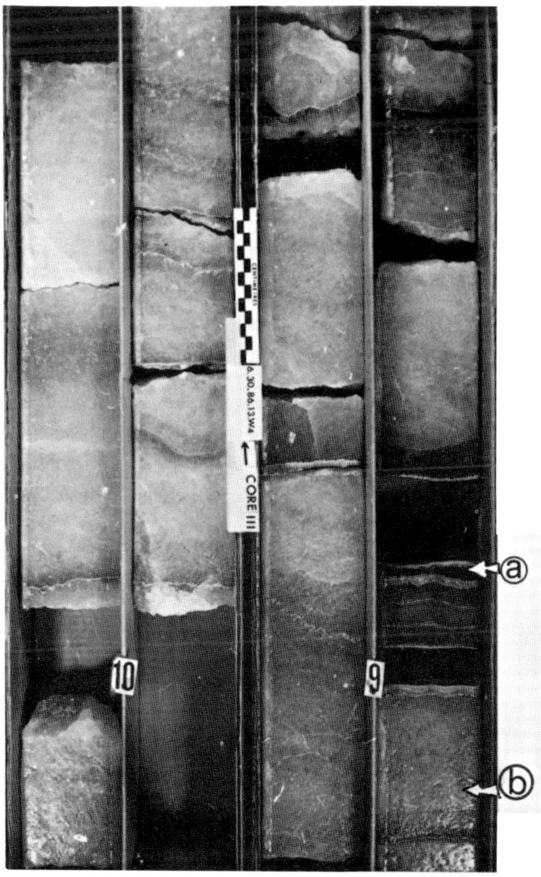

2

PLANCHE 22 SONDAGE 6.30.86. 13 W4 (suite) : SECTIONS DE CAROTTES, CONTACT SEL/CARBONATES

Fig. 1 et 2 Sections de carottes du sel de fosse.

Fig. 3 Contact sel/carbonates.

Cette planche est destinée à illustrer le "sel de fosse". Rappelons d'abord les caractères de ce "sel de fosse" qui apparaissent sur les diagraphies (planche 20) : non seulement il n'existe pratiquement aucune intercalation non salifère d'une épaisseur appréciable par rapport au pas des outils, intercalation qui se serait marquée par un pic sur les différents enregistrements ; mais en outre la courbe très plate du gamma ray suggère une grande homogénéité confirmée par l'allure et la valeur de la courbe densité (qui indique en même temps une grande pureté).

Les carottes illustrées se situent à deux niveaux de cette masse salifère : la base du premier niveau (base de la boîte 10) est à 10 m environ au-dessus de la base du sel ; le sommet du deuxième niveau (sommet de la boîte 1) n'est qu'à 3 m du toit de ce sel. Notons que les deux niveaux sont séparés par une dizaine de mètres. La similitude d'aspect de ces niveaux apparaît donc remarquable et représentative.

Le sel massif est pur, blanc, translucide et cristallin, le développement cristallin étant toutefois moindre que dans le "sel de nappe" illustré sur la planche suivante. Les impuretés se groupent en niveaux grisâtres, ternes, d'épaisseur millimétrique ou centimètrique, bien stratifiés. Il s'agit d'un matériel limoneux, carbonaté, voire argileux (?). Dans les carottes illustrées, ces intercalations d'insolubles deviennent plus abondantes dans la boîte 9 : une figure de détail (fig. 3) montre le haut du sel et une partie de ces intercalations carbonatées et l'allure du contact qui les sépare. L'allure ondulée de la surface supérieure du sel fait évoquer les *ripple marks* signalés par exemple par Delwig dans la formation Salina du Michigan (New York) également au contact de sel et d'insolubles ; comme si les eaux, ayant ramené des insolubles dans le bassin, avaient en arrivant **remanié et remodelé** la surface supérieure du sel. Sur une simple section de carotte, il est difficile d'en dire plus, en particulier sur l'apparente symétrie de ces ondulations, symétrie théoriquement mal compatible avec une interprétation de *ripple marks*. Dans les creux de ces ondulations, on remarquera l'allure du remplissage laminé, très fin, de nature carbonatée. On notera en outre que les ondulations du sel sous-jacent s'amortissent rapidement dans les laminations qui affectent l'intercalation carbonatée.

Au-dessus de cette première intercalation carbonatée, de nouveaux niveaux carbonatés en général plus fins vont alterner avec des lits d'un sel grisâtre, le tout étant bien interstratifié. De discrètes irrégularités sédimentaires apparaissent encore dans une de ces zones salifères : un plissotement d'origine non précisée est encadré par des lits plus calmes.

Les observations faites sur les plissottements qui affectent la série holocène du Zuiderzee (Van Loon, A.J. and Wiggers, A.J., 1976)([1]) montrent que ces phénomènes peuvent naître sans aucune pente ni déformation tectonique.

([1]) Abrasion as an agent for sand supply in a Holocene lagoon (Almere and Zuiderzee Members, Groningen formation) in the Netherlands. *Sediment. Geol.*, **15**, 293-307.

PLATE 22 BOREHOLE 6.30.86.13 W4 (continued) : CORE SECTIONS, SALT/ CARBONATES CONTACT

Figs. 1 and 2 Trough salt slabed cores.

Fig. 3 Salt/carbonates boundary.

This plate aims to show "trough salt". The features of this "trough salt" as shown on the logs (Plate 20) are worth recalling : not only is there hardly any non-saliferous interbedding of reasonable thickness with regard to the detection limit, an interbedding which would be marked by a peak on the different records, but also the flat gamma ray curve suggests considerable homogeneity, confirmed by the appearance and values of the density curve (which, at the same time, indicates high purity).

The cores shown are from two different layers of this saliferous mass : the base of the first layer (bottom of box 10) is about 10 m above the bottom of the salt ; the top of the second layer (top of box 1) is only 3 m above the roof of the salt. The two layers are separated by about 10 m. The similarity between the two layers is thus noticeable and significant.

The massive salt is pure, white, translucent and crystalline, the crystalline development being nevertheless less than in the "sheet salt" seen in the next plate.

Impurities collect in well-bedded greyish layers, of millimetric or centimetric thickness. They are carbonated, even argillaceous (?) silts. In the illustrated cores, these insoluble layers become more common in box 9 : a detailed view (Fig. 3) shows the top of the salt and some of the carbonated parts as well as the limit between the two. The undulation at the upper surface of the salt recalls the "riple marks", mentioned by Delwig for example, in the Salina formation of Michigan, New York, at the boundary between the salts and the insolubles suggesting that the waters brought the insolubles down into the basin had, once there, reworked and remodelled the salt's upper surface. In a simple core section it is not easy to say anymore, especially on the apparent symmetry of the undulations, a symmetry theoretically incompatible with a ripple mark interpretation. In the troughs between these undulations, laminated very fine carbonated in-filling can be seen. Furthermore, we note that the underlying salt undulations weaken greatly in the laminations affecting the carbonated intercalation.

Just above this level, new generally thinner carbonated levels start to alternate a greyish salt, both being well interbedded. Slight sedimentary irregularities continue to occur in one of these saliferous zones : a crumpling of undefined origin is surrounded by calmer beds.

Observations made on crumplings affecting the holocene series of Zuiderzee (Van Loon and Wiggers, 1976) ([1]) show that these phenomena may come about without any slope or tectonic distortion.

([1]) Abrasion as an agent for sand supply in a Holocene lagoon (Almere and Zuiderzee Members, Groningen formation) in the Netherlands. *Sediment. Geol.*, **15**, 293-307.

PLANCHE 23 SONDAGE 6.30.86.13 W4 (suite et fin) : SECTIONS DE CAROTTES ET SEL CRISTALLIN ASSOCIE A DE LA BOUE CARBONATEE

(Voir texte de la légende page 128).

PLATE 23 BOREHOLE 6.30.86.13 W4 (continued to end) : CORE SECTIONS AND CRYSTALLINE SALT ASSOCIATED WITH CARBONATED MUD

(See text of legend page 129).

1

2

PLANCHE 23 SONDAGE 6.30.86.13 W4 (suite et fin) : SECTIONS DE CAROTTES ET SEL CRISTALLIN ASSOCIE A DE LA BOUE CARBONATEE

Fig. 1 Sections de carottes dans le sel de nappe.

Fig. 2 Sel cristallin et boue carbonatée.

Le "sel de nappe", à l'inverse du "sel de fosse", est caractérisé par une très vaste extension régionale. Chacune des unités de "sel de nappe" évolue très lentement (en ce qui concerne son épaisseur et le pourcentage de halite) alors que le "sel de fosse" reste partout très dur et se biseaute rapidement sur les bords de la fosse. Les diagraphies de la planche 20 montrent déjà des caractères pétrographiques qui diffèrent de ceux du sel de fosse : allure plus hachée des différentes courbes, traduisant une moindre pureté (notamment sur les logs densité) et intercalations d'insolubles suffisamment épaisses pour se traduire par des pics bien caractérisés (voir logs de la planche 20). Les figures de la planche ci-jointe précisent des traits pétrographiques à plus grande échelle.

Le sel gemme transparent, hyalin, de couleur ambrée, apparaît moins "propre" que le "sel de fosse" de la planche précédente. Par contre il est mieux cristallisé : les cristaux centimétriques présentent parfois des formes cubiques presque automorphes, avec de très nets clivages. Il est intéressant de noter la présence d'une boue carbonatée, micritique, non laminée, souvent concentrée à la limite des faces du sel cristallin (voir figure d'ensemble et figure de détail). Ce mode de gisement semble pouvoir s'interpréter par des phénomènes successifs suivants : **dissolution** du sel, puis au fur et à mesure d'une phase de **croissance cristalline**, la boue carbonatée présente se trouve repoussée et reconcentrée en ces filets micritiques. Le processus global n'est pas sans évoquer, suivant A. Maurin, la stylolitisation : par transfert de solubilités, le sel jouerait ici le rôle de calcaire dans la stylolitisation tandis que les carbonates de la série salifère joueraient le rôle des insolubles des calcaires à stylolites.

On notera en outre, par exemple à la base de la boîte 6, la présence d'insolubles sous la forme d'intercalations stratifiées comprenant carbonates et, semble-t-il à ce niveau, nodules d'anhydrite.

Les carbonates peuvent encore être finement laminés et présenter des figures de plissement (glissement ou débuts de chevauchements ?). De telles intercalations seraient responsables des pics observés sur diagraphies.

PLATE 23 BOREHOLE 6.30.86.13 W4 (continued to end) : CORE SECTIONS AND CRYSTALLINE SALT ASSOCIATED WITH CARBONATED MUD

Fig. 1 Sheet salt slabed cores.

Fig. 2 Crystalline salt and carbonated mud.

The "sheet salt" contrary to the "trough salt", is very widespread over the region. Each of the "sheet salt" units evolves very slowly (as regards its thickness and halite content), whereas the "deep salt" remains very pure throughout and bevels rapidly on the edges of the basin. The logs of plate 20 show petrographical features which differ from those of the "deep salt" : the different curves are more discontinuous, showing lesser purity (especially in density logs) ; intercalations of insolubles sufficiently thick to be shown as characteristic deflections (see logs on Plate 20). The figures on the accompanying plate show petrographic features on a larger scale.

The clear amber coloured rock salt, is less "clean" than the "trough salt" in the previous plate. It is, on the other hand, better crystallized. The centimetric crystals are sometimes cubic, nearly automorphic and with very perfect cleavages. One should note the presence of micritic, non laminated carbonated mud often concentrated at the edge of crystalline salt facies (see general figure and detailed view). It seems possible to interpret this kind of deposit using the following successive phenomena : salt solution, then during a period of crystalline growth the carbonated mud becomes pushed back and rearranged in micritic streamlets. According to A. Maurin, the total process can evoke stylolitisation. Salt here would play the role of limestone in stylolitisation while the carbonates of the saliferous series would play here the role of insoluble organic matter in stylolitisation.

We may also note, at the base of box 6 for example, the presence of bedded insoluble layers, including carbonates and what appear to be anhydrite nodules.

The carbonates may still be finely laminated, and show folding (slumping or the beginnings of thrust faulting ?). Such intercalations would be the cause of the peaks observed in the logs.

CHAPITRE IV (Planches 18 à 30). *CHAPTER IV (Plates 18 to 30)*

Successions carbonatées, anhydritiques et salifères dans le Dévonien moyen de forages du nord-est de l'Alberta (Canada occidental).

Carbonated, anhydritic and saliferous sequences in the Middle Devonian of wells in North-Eastern Alberta (West Canada).

PLANCHE 24 SONDAGE 16.35.115.2 W5 : SECTIONS DE CAROTTES ET LAMINITES

Fig. 1 et 2 Sections de carottes de la base du Muskeg.

Fig. 3 Laminites bréchifiées.

Fig. 4 Laminites carbonatées.

La zone illustrée dont la base se trouve sur la planche ci-jointe débute environ 2 m au-dessus du contact Keg River-Muskeg (défini par l'apparition du premier banc d'anhydrite). A la base sur 2,20 m environ (épaisseur minimale puisque limitée par la base de la boîte 2 de la carotte 4 étudiée) s'observent des laminites carbonatées, très fines (d'ordre millimétrique), plan-parallèles, parfois faiblement ondulées. Sur la figure de détail (fig. 4), on peut voir que les bancs carbonatés clairs sont légèrement plus épais que les bancs sombres qui les séparent.

Après ces laminites, de l'anhydrite occupe actuellement environ 5,40 m d'épaisseur de séries (boîte 1, carotte 4 de la planche ci-jointe ; boîtes 14, 13, 12 en partie, de la carotte 3, illustrées sur les planches 25 et 26). Sur les 1,40 m environ ici présents, l'anhydrite apparaît massive, à structure mosaïque, de couleur claire. Elle emballe des fragments assez **anguleux** et tordus de carbonates laminés dont l'un a été agrandi en bas à droite (fig. 3). Les laminations ont été ployées, elles ont glissé et se sont cassées, ce qui implique qu'une partie des mouvements ait eu lieu après induration. Une certaine digestion apparaît à la base de l'image agrandie où des résidus de laminites "flottent" dans une matrice anhydritique. Enfin dans le cœur des figures de glissement (*slump* ?) de l'anhydrite blanche secondaire s'est développée par remaniement, dissolution, et cristallisation successifs.

PLATE 24 BOREHOLE 16.35.115.2 W5 : CORE SECTIONS AND LAMINITES

Figs. 1 and 2 Lower part of the Muskeg, slabed cores.

Fig. 3 Brecciated laminites.

Fig. 4 Carbonated laminae.

The zone shown here, the basal part of which is illustrated on this plate, begins about 2 m above the Keg River-Muskeg boundary (marked by the occurrence of the first anhydrite bed). At the base, over about 2.20 m (minimal thickness, being limited by the base of box 2 from core 4) carbonate laminites are seen which are very fine (millimetric), in plan parallel arrangement and sometimes slightly undulating. In the detailed view photo (Fig. 4), the light carbonate beds are slightly thicker than the darker beds which separate them.

After these laminites, anhydrite then constitutes 5.40 m (box 1, core 4 in the adjoining plate ; boxes 14, 13 and 12 in part : core 3, illustrated in plates 25 and 26). Over about 1.40 m light coloured, mosaic massive anhydrite occurs, surrounding quite **angular,** twisted fragments of laminated carbonates (see magnification, Fig. 3). The laminations are bent, slumped and broken, suggesting that some of these movements occured after induration. At the base of the magnification, a digestion can be seen where laminite remains "float" in an anhydritic matrix. Finally, in the center of the slump structure, white secondary anhydrite has developed by successive reworking, dissolution and crystallization.

1

a

1635 115 2W5 IV

b

2

a

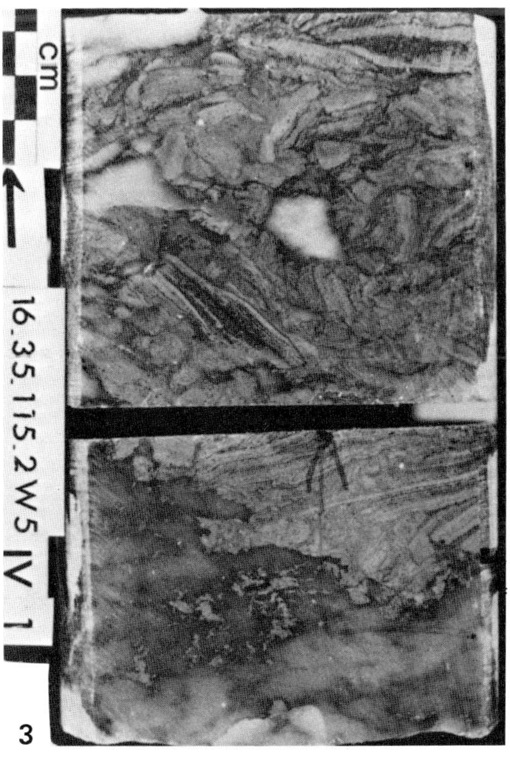

3

b

4

IV.2

CHAPITRE IV (Planches 18 à 30). *CHAPTER IV (Plates 18 to 30)*

Successions carbonatées, anhydritiques et salifères dans le Dévonien moyen de forages du nord-est de l'Alberta (Canada occidental).

Carbonated, anhydritic and saliferous sequences in the Middle Devonian of wells in North-Eastern Alberta (West Canada).

PLANCHE 25 SONDAGE 16. 35. 115. 2 W5 (suite) : SECTIONS DE CAROTTES ET PLISSEMENT DANS L'ANHYDRITE

Fig. 1 Sections de carottes du Muskeg.

Fig. 2 Plissement *in situ* dans l'anhydrite.

La carotte 3, dont les boîtes basales sont illustrées ici, est en continuité avec la carotte 4 précédente.

Les deux boîtes (14 et 13 de la planche ci-jointe), soit environ 3,60 m, sont occupées par l'anhydrite massive déjà évoquée. On notera, à la base de la colonne droite de la boîte 14, la présence, au milieu d'une anhydrite mosaïque sombre, de passées à structures entérolithiques de couleur plus claire. Les fragments carbonatés s'y retrouvent, surtout dans la première moitié de la boîte 14 : fragments de quelques centimètres, laminés, avec figures de glissement. Vers le sommet de l'intervalle (boîte 13 et figure de détail 2), une passée entérolithique d'anhydrite grise est soulignée par un liseré, non continu, d'impuretés vraisemblablement carbonatées. L'hypothèse d'un plissement *in situ* apparaît plus vraisemblable ici que la coalescence de nodules pour expliquer cette structure entérolithique. La figure de plissement est bien marquée par un lit contourné d'anhydrite blanchâtre, très pure, traduisant semble-t-il une évolution sédimentaire et non diagénétique. Un nouveau cycle démarre au-dessus de l'anhydrite blanche par un épisode "sale" à impuretés organiques.

PLATE 25 BOREHOLE 16. 35. 115. 2 W5 (continued) : CORE SECTIONS AND FOLDING UP IN THE ANHYDRITE

Fig. 1 Muskeg slabed cores.

Fig. 2 *In situ* anhydrite folding.

Core 3, whose lower boxes are shown here, continues from core 4.

The two boxes (14 and 13 in the accompanying plate), about 3.60 m, are filled with massive anhydrite, as already mentioned. At the base of the right hand column of box 14, we may observe, within dark coloured, mosaic anhydrite, the presence of lighter coloured layers of enterolithic structure. Carbonate fragments are again present, particularly in the first half of box 14 : laminated fragments of several centimeters, with slump structures. Towards the top of the interval (box 13 and detailed Fig. 2), an enterolithic layer of grey anhydrite is emphasized by a broken border of impurities, which are probably carbonate. The suggestion of an *in situ* folding seems a more likely explanation for this enterolithic structure than the coalescence of nodules. The folding is emphasized by a very pure distorted white anhydritic bed of a sedimentary, as opposed to diagenetic, evolution. A new cycle starts above the white anhydrite with a "dirty" episode containing organic impurities.

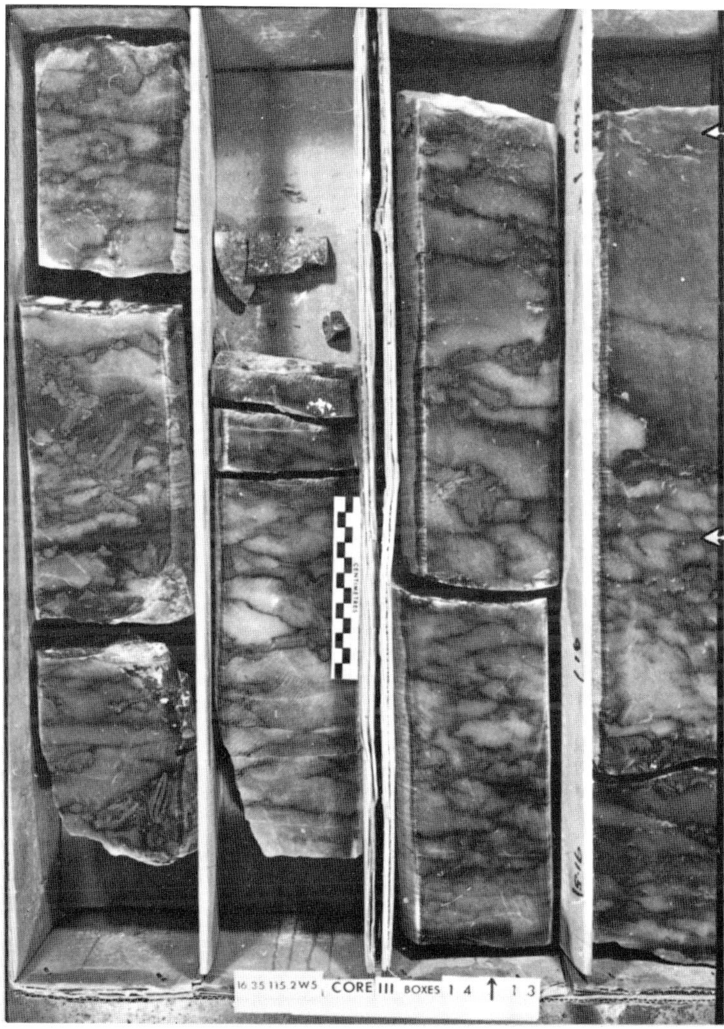
16 35 115 2W5 CORE III BOXES 1 4 ↑ 1 3

1

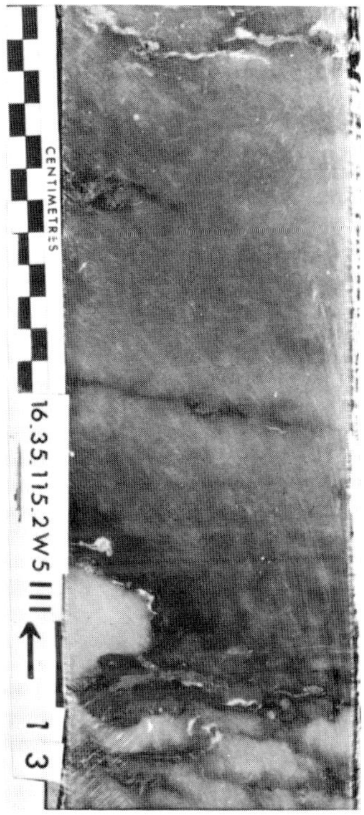
CENTIMETRES
16.35.115.2W5 III ← 1 3

2

PLANCHE 26 SONDAGE 16.35.115.2 W5 (suite) : SECTIONS DE CAROTTES, DOLOMIES LITEES ET LAMINEES

(Voir texte de la légende page 136).

PLATE 26 BOREHOLE 16.35.115.2 W5 (continued) : CORE SECTIONS, BEDDED AND LAMINATED DOLOMITES

(See text of legend page 137).

a
b
c
16.35.115.2W5 CORE III BOXES 1 2 ↑ 11

a

2

16.35.115.2W5
III
1 2

b

c

16.35.115.2W5
III
1 2

3

4

CHAPITRE IV (Planches 18 à 30). *CHAPTER IV (Plates 18 to 30)*

Successions carbonatées, anhydritiques et salifères dans le Dévonien moyen de forages du nord-est de l'Alberta (Canada occidental).

Carbonated, anhydritic and saliferous sequences in the Middle Devonian of wells in North-Eastern Alberta (West Canada).

PLANCHE 26 SONDAGE 16.35.115.2 W5 (suite) : SECTIONS DE CAROTTES, DOLOMIES LITEES ET LAMINEES

Fig. 1 Sections de carottes du Muskeg.

Fig. 2 Dolomie laminée plan-parallèle.

Fig. 3 Dolomie litée avec plissotements et glissements.

Fig. 4 Dolomie litée à fins lits noirs.

Sur les 45 premiers centimètres de la boîte 12, l'anhydrite massive se poursuit, avec quelques filets carbonatés. Une zone de transition d'environ 1 cm où se mêlent anhydrite et carbonate marque le passage de l'unité suivante.

Le reste de la boîte 12 et la première moitié de la boite 11 (soit environ 2,25 m) sont occupés par une dolomie litée. Celle-ci présente en effet une stratification d'une épaisseur d'ordre plurimillimétrique à centimétrique (le terme de lamination devant être réservé à une périodicité millimétrique ou inframillimétrique). Le premier échantillon agrandi (fig. 3), situé légèrement au-dessus de l'anhydrite massive, offre de beaux exemples de phénomènes de plissottement, de glissement, pouvant aller jusqu'à la pseudo-nodulisation. Insistons sur la localisation, une fois encore, de ces indices de mouvements importants, au toit d'une épaisse couche de sulfate.

La deuxième colonne de la boîte 12 montre un plissement d'amplitude presque métrique, mis en évidence à la fois par l'inversion des pendages et surtout par la présence d'un coeur de pli couché (vers la moitié de la hauteur). Du fait de ce plissement, une partie des couches se présente donc dans un ordre de succession renversée. Lithologiquement, les couches affectées sont plus finement litées que celles de la colonne précédente. A la partie supérieure, principalement, de fins lits noirs correspondent à l'exsudation de bitume à partir d'un matériel dolomitique qui se trouvait donc doué d'une certaine porosité (détail fig. 4). Rappelons que des couches de cette nature ont été souvent interprétées, il y a un peu moins d'une décennie, comme étant de nature stromatolithique ; interprétation qu'aucun fait d'observation n'a jamais confirmé.

La première moitié de la boîte 11 (figure d'ensemble et figure de détail 2) consiste encore en une dolomie où le litage, continuant à s'affiner, passe à une lamination plan-parallèle, faisant alterner couches claires et couches sombres riches en matière organique. Certains lits noirs semblent, cependant, avoir encore pour origine l'exsudation de bitume. La pente de ces couches, en continuité avec l'important plissement observé dans la boîte précédente, pourrait donc avoir la même origine, sans rapport éventuel avec un pendage régional.

Dans les 20 derniers centimètres de la boîte 11, la dolomie, perdant ses lits noirs, devient homogène et poreuse, finement cristalline, sans éléments figurés remarquables macroscopiquement.

PLATE 26 BOREHOLE 16.35.115.2 W5 (continued) : CORE SECTIONS, BEDDED AND LAMINATED DOLOMITES

Fig. 1 Slabed cores of the Muskeg.

Fig. 2 Plan-parallel laminated dolomite.

Fig. 3 Bedded dolomite with crumplings and slumpings.

Fig. 4 Bedded dolomite with thin dark layers.

Massive anhydrite continues in the first forty five centimeters of box 12 with some carbonate thin strata. A transitional zone of about 1 cm with a mixture of anhydrite and carbonate leads to the next unit.

The rest of box 12 and the first half of box 11 (about 2.25 m) are made up of bedded dolomite which has a stratification several millimeters to a centimeter thick (the term "lamination" should only be applied to a millimetric or inframillimetric range). The first sample, see wider magnification (Fig. 3), which lies just above the massive anhydrite bears good examples of crumpling and slumping up to pseudo-nodulisation. The localisation of these indices of large scale movements on the top of a thick sulfate layer is again worth noting.

The second column of box 12 shows folding of nearly metric amplitude, indicated by the inversion of dips and particularly by the presence of a recumbent fold core (about half way up). Because of this folding, some of these beds are seen in a reversed order of succession. From the lithological point of view, the layers affected are more finely bedded than those of the previous column. Mainly in the upper portion, fine black layers mark bitumen exudation from dolomitic material which was reasonably porous (detailed view, Fig. 4). A little less than a decade ago, layers of this type were often interpreted as stromatolithic, but this has never been confirmed by observations.

The first part of box 11 (figure of the whole and detailed view 2) is again of dolomite, in which the bedding, becoming thinner, progresses to plan parallel lamination with an alternation of light layers and dark ones, rich in organic material. However, some of the black layers still seem to originate from bitumen exudation. The dip of these layers, in continuity with the considerable folding seen in the previous box, may therefore have the same and may possibly have no bearing on the regional dip.

In the last 20 cm of box 11, the black beds disappear from the dolomite, which becomes homogeneous, porous, finely crystalline and has no macroscopically noticeable elements.

PLANCHE 27 SONDAGE 16.35.115.2 W5 (suite) : SECTIONS DE CAROTTES ET DOLOMIE LITEE A PLISSEMENT

Fig. 1 Sections de carottes du Muskeg.

Fig. 2 Dolomie finement litée à plissements.

Les 45 premiers centimètres de la boîte 10 correspondent à la suite de cette dolomie homogène et poreuse.

Le reste de la boîte 10 et la boîte 9 (soit environ 3 m, se poursuivant d'ailleurs dans les boîtes 8 et 7) sont occupés par des alternances de dolomies finement litées et laminées d'une part, et, d'autre part, de dolomies homogènes. Ces dernières se présentent en passées d'ordre décimétrique, et participent à un peu moins de la moitié de l'épaisseur totale. Les passées laminées sont encore parfois affectées d'un certain pendage. Mais surtout, des plissements très accusés rebroussent très fortement, sans cassure ni rupture importantes, des zones de quelques centimètres, néanmoins encadrées par des laminites parfaitement plan-parallèles (boîte 10 et figure de détail 2). De tels phénomènes sont fréquents, dans le bassin, puisque Davies et Ludlam (1973) en ont illustré, de leur côté, de très beaux exemples, dans d'autres coupes de ce Muskeg.

PLATE 27 BOREHOLE 16.35.115.2 W5 (continued) : CORE SECTIONS AND BEDDED DOLOMITE WITH FOLDING

Fig. 1 Slabed cores of the Muskeg.

Fig. 2 Finely bedded dolomite with foldings.

The first 45 cm of box 10 show a continuation of this homogeneous, porous dolomite.

The rest of box 10 and box 9 (about 3 m and also boxes 8 and 7) contain, on the one hand, finelv bedded dolomites and laminated dolomites in alternation, and on the other, homogeneous dolomites. The latter are represented in decimetric layers over slightly less than half of the total thickness. The laminated layers are still sometimes affected by some dip. But, more importantly, very obvious folds markedly upturn zones several centimetres thick, without entailing great breaks or ruptures, although these are surrounded by perfectly plan parallel beds of laminites (box 10 and detailed view 2). Such phenomena are commonly encountered in the basin ; Davies and Ludlam (1973) have, for their part, described some very good examples in other sections from the Muskeg.

16.35.115.2W5, CORE III BOXES 10 ↑ 9

1

CENTIMETRES
16.35.115.2W5 III 10

2

CHAPITRE IV (Planches 18 à 30). *CHAPTER IV (Plates 18 to 30)*

Successions carbonatées, anhydritiques et salifères dans le Dévonien moyen de forages du nord-est de l'Alberta (Canada occidental).

Carbonated, anhydritic and saliferous sequences in the Middle Devonian of wells in North-Eastern Alberta (West Canada).

PLANCHE 28 SONDAGE 16.35.115.2 W5 (suite) : SECTIONS DE CAROTTES ET ALTERNANCES DOLOMIE HOMOGENE-DOLOMIE LAMINEE

Fig. 1 et 3 Sections de carottes du Muskeg.

Fig. 2 et 4 Détails d'alternances dolomie homogène-dolomie laminée.

Sur ces deux boîtes, le régime d'alternances dolomie homogène-dolomie laminée se poursuit. On notera la présence de laminations d'une finesse et d'une régularité extraordinaires (boîte 7 par exemple et photos de détail). Quelques figures de glissement y sont encore visibles. Les passées non laminées offrent elles-mêmes ici une différenciation : les premiers centimètres, ceux de la base, sont nettement moins homogènes, présentent des filets sombres, en chevelu, de la pyrite finement exprimée (fig. 4). Au-dessus, les 1 ou 2 cm précédant la réapparition des laminites sont caractérisés par une structure beaucoup plus finement homogène, d'aspect estompé (voir les deux photos de détail). Dans la dernière passée non laminée (50 cm sommitaux de la boîte 7), la dolomie peu homogène contient de l'anhydrite en abondance, sous la forme de baguettes et de rosettes dispersées, trop petites pour être visibles sur le cliché ci-joint.

Enfin la stylolitisation prend ici une ampleur exceptionnelle et l'absence d'altération, habituelle aux affleurements, lui confère un aspect spectaculaire. En outre, la noirceur des éléments insolubles rend très visible le front de stylolitisation. Il serait imprudent de vouloir schématiser la position de ces phénomènes, par exemple au contact des passées homogènes et des passées laminées. Il peut exister deux générations de stylolitisation (fig. 3), le front de la plus récente ayant oblitéré une partie de la plus ancienne.

Les discordances angulaires créées par certains stylolithes (fig. 2) pourraient aboutir, par empilement ou répétition, à la création d'un faux pendage.

PLATE 28 BOREHOLE 16.36.115.2 W5 (continued) : CORE SECTIONS AND ALTERNATIONS OF HOMOGENEOUS DOLOMITE-LAMINATED DOLOMITE

Figs. 1 and 2 Slabed cores of the Muskeg.

Figs. 3 and 4 Details of alternations homogeneous dolomite and laminated dolomite.

The alternation of homogeneous and laminated dolomite continues in these two boxes. The presence of extraordinarily fine and regular laminations should be noted (box 7, for example, and detailed views). Some slumping forms can still be seen here, but the non-laminated passages mark a change : the first centimetres at the base are clearly less homogeneous, with fine, dark anastomosed pyrite strata (Fig. 4). Just above, in the 1 or 2 cm which precede the reappearance of laminites, the structure is much more homogeneous, with a blurred appearance (see the two detailed views). In the last non-laminated passage (the top 50 cm of box 7) unhomogeneous dolomite contains abundant anhydrite in dispersed laths and rosettes, too small to be distinguished in the attached photograph.

Finally, the scale of stylolitisation here is quite spectacular, and especially so because of the lack of alteration which is normally expected in outcrops. Furthermore, the blackness of insoluble elements emphasizes the stylolitisation limits. One should not try to schematise the position of these features, such as the boundary between homogeneous and laminated passages. Two generations of stylolitisation could possibly exist (Fig. 3), the more recent one having obliterated some of the previous one.

The angular unconformities caused by certain stylolits (Fig. 3) could, if they were piled up or repeated, lead to the creation of an artificial dip.

a

b

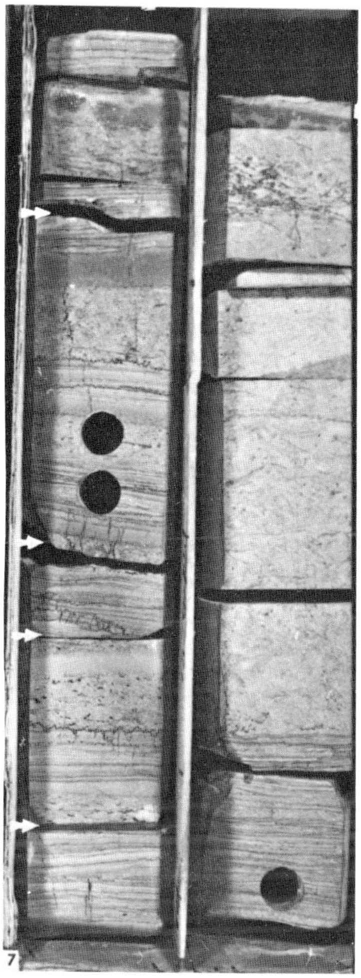
7

1

16 35 115 2W5
CORE III BOXES 8

2

a

CENTIMETRES
16.35.115.2W5
III 7

3

b

16.35.115.2W5
III
7

4

CHAPITRE IV (Planches 18 à 30). *CHAPTER IV (Plates 18 to 30)*

Successions carbonatées, anhydritiques et salifères dans le Dévonien moyen de forages du nord-est de l'Alberta (Canada occidental).

Carbonated, anhydritic and saliferous sequences in the Middle Devonian of wells in North-Eastern Alberta (West Canada).

PLANCHE 29 SONDAGE 16.35.115.2 W5 (suite) : SECTIONS DE CAROTTES ET LAMINATIONS.

Fig. 1 Sections de carottes du Muskeg.

Fig. 2 Laminations soulignées par de l'anhydrite en lits.

Fig. 3 Laminations à plissotements et plissements.

Fig. 4 Laminations irrégulières avec lentilles d'anhydrite.

Une lacune (boîtes 6 et 5) sépare la coupe illustrée sur la planche précédente de la partie ici représentée.

Les 90 premiers centimètres consistent en dolomies où le litage n'est que peu apparent. Des exsudations de bitume à la partie moyenne proviennent d'une porosité laminaire et s'étalent sur la carotte.

Le mètre suivant (deuxième moitié de la boîte 4 et base de la boîte 3) est représenté par une série laminée. De remarquables plissottements et plissements (figure de détail 3, en bas à gauche) rappellent de façon frappante ceux qui ont été illustrés dans les planches antérieures (planche 27, par exemple). La lamination de ces couches est moins régulière, moins plan-parallèle que celle existant quelques mètres plus bas (boîte 7 par exemple).

Sur 1,70 m environ (boîte 3 *p.p. max.*), la lamination et le litage deviennent moins fins et moins réguliers, en même temps que réapparaît l'anhydrite exprimée macroscopiquement. Cette dernière peut se présenter en yeux et en lentilles (figure de détail 4), ou en lits centimétriques traversant toute la carotte (par exemple figure de détail 2). En outre, les lits sombres, responsables du litage des carbonates, tendent à perdre leur continuité, leur régularité, et leur parallélisme (allure "chaotique" dans le litage).

PLATE 29 BOREHOLE 16.35.115.2 W5 (continued) : CORE SECTIONS AND LAMINATIONS

Fig. 1 Slabed cores of the Muskeg.

Fig. 2 Laminations with bedded anhydrite.

Fig. 3 Laminations with crumplings and foldings.

Fig. 4 Irregular laminations with lenses of anhydrite.

A gap (boxes 6 and 5) separates the section shown on the previous plate from the part seen here.

The first 90 cm are made up of dolomites in which there is only slight bedding. Bitumen exudations in the middle part are caused by laminar porosity and are spread throughout the core.

The following metre (second half of box 4 and bottom of box 3) consists of a laminated serie. Considerable crumplings and foldings (detailed view, Fig. 3) are striking reminders of the phenomena manifested in previous plates (Plate 27, for example). The lamination of these layers is less regular and less parallel than it is several metres lower down (box 7, for instance).

Over about 1.70 m (box 3 *p.p. max.*), lamination and bedding become less fine and more irregular and anhydrite reoccurs macroscopically. This anhydrite may occur in eyes or lenses (detailed view, Fig. 4) or in centimetric beds across the whole core (for example, detailed view, Fig. 2). Furthermore, the dark beds responsible for carbonate bedding tend to lose their continuity, regularity and parallel arrangement ("chaotic" character in the bedding).

2

a

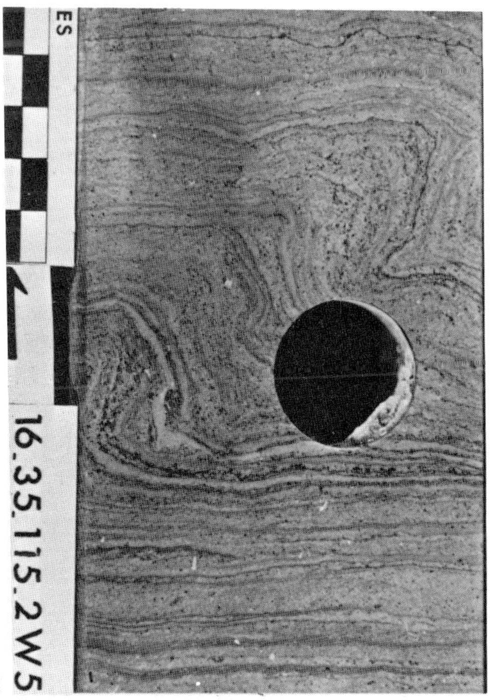

c

3

b

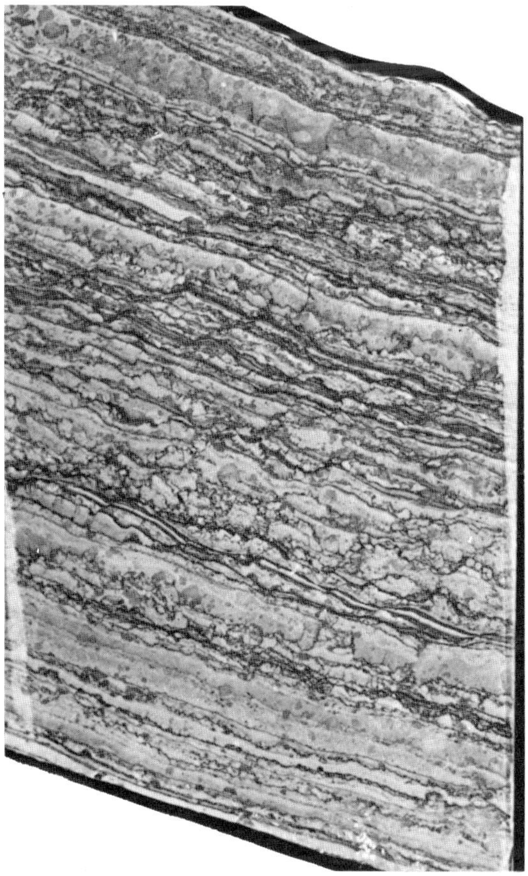

4

CHAPITRE IV (Planches 18 à 30). *CHAPTER IV (Plates 18 to 30)*

Successions carbonatées, anhydritiques et salifères dans le Dévonien moyen de forages du nord-est de l'Alberta (Canada occidental).

Carbonated, anhydritic and saliferous sequences in the Middle Devonian of wells in North-Eastern Alberta (West Canada).

PLANCHE 30 SONDAGE 16.35.115.2 W5 (suite et fin) : SECTIONS DE CAROTTES, ANHYDRITE ENTEROLITHIQUE ET LAMINITES

Fig. 1 Anhydrite entérolithique sédimentaire.

Fig. 2 et 3 Section de carottes du Muskeg.

Fig. 4 Laminites plissotées et laminites planes.

1,80 m environ (boîte 2 de la carotte 3) de dolomie à laminations plus ou moins régulières et plus ou moins fines, où l'anhydrite disséminée n'est plus visible macroscopiquement.

Sur 1,40 m (boîte 1) alternent des passées décimétriques de dolomies peu laminées à anhydrite disséminée et des passées laminées. Des plissottements, glissements (*slumps*) affectent ici encore des zones de quelques centimètres de laminites, laissant remarquablement indemnes de toute déformation les laminites immédiatement sus-jacentes (base de la deuxième moitié de la boîte 1, sur la photo d'ensemble 3, figure de détail 4).

Sur 0,60 m environ (sommet de la boîte 1), avec anhydrite finement dispersée, non visible macroscopiquement : deux alternances superposent des lits pluricentimétriques sombres (laminés) et des lits de dolomie claire homogène. Le lit sombre du bas est en fait redoublé par un plissottement qui semble avoir déclenché le pendage visible au-dessous (deuxième lit sombre).

2,70 m (boîte 14 et première moitié de la boîte 13 de la carotte 2) de laminites carbonatées et anhydritiques : alternances de laminations millimétriques, plan-parallèles, sub-horizontales ou peu inclinées et de lits d'anhydrite pouvant atteindre 1 cm d'épaisseur, en lits continus sur toute la carotte ou en lentilles responsables d'un aspect ocellé ou boudiné. Ces alternances sont affectées par des plissottements où l'anhydrite peut finir par affecter une structure entérolithique "chevauchante" (figure de détail 1). L'allure de ces alternances et le plissottement des deux constituants, établissent qu'il s'agit là d'anhydrite sédimentaire et non de dépôt diagénétique (par remplacement ou par déplacement).

0,90 m environ (épaisseur minimale : fin de la zone illustrée) d'alternances pluricentimétriques de dolomies à fines laminations plan-parallèles, sub-horizontales, riches en petits nodules d'anhydrite, non visibles macroscopiquement et de dolomie homogène non laminée.

PLATE 30 BOREHOLE 16.35.115.2 W5 (continued to end) : CORE SECTIONS, ENTEROLITHIC ANHYDRITE AND LAMINITES

Fig. 1 Sedimentary enterolithic anhydrite.

Figs. 2 and 3 Slabed cores of the Muskeg.

Fig. 4 Crumpled laminites and plane laminites.

About 1.80 m (box 2 of core 3) of dolomite with laminations which are more or less regular and fine and where the scattered anhydrite is no longer macroscopically visible.

Over 1.40 m (box 1) decimetric intercalations of poorly laminated dolomites with scattered anhydrite alternate with laminated passages. Crumpling and slumping still affect zones with several centimetres of laminites, leaving immediately overlying laminites amazingly undamaged by any distortion (bottom of second half of box 1, in the photo of the whole, Fig. 3 and the detailed view, bottom right, Fig. 4).

Over about 0.60 m (top of box 1), with finely distributed anhydrite not macroscopically visible : dark (laminated) pluricentimetric beds alternate twice with homogeneous, light coloured dolomite beds. The dark lower bed is in fact doubled by a crumpling, which seems to have initiated the dip seen below (second dark bed).

2.70 m (box 14 and first half of box 13 of core 2) of carbonate and anhydritic laminites : alternations of millimetric, parallel-lying, sub-horizontal or scarcely sloping laminations with beds of anhydrite up to 1 cm thick in beds continuing throughout the whole core, or in lenses which cause the ocellate or slubed appearance. These alternations are disturbed by crumpling, where anhydrite may eventually cause an enterolithic, "overlapping" structure (detailed view, top left). The character of these alternations and the crumpling of the two constituents prove that the anhydrite is of sedimentary, not diagenetic deposition (by replacement or displacement).

About 0.90 m (minimal thickness : end of illustrated zone) of pluricentimetric alternations of dolomites with fine, parallel and sub-horizontal laminations, which are rich in small anhydrite nodules not macroscopically visible, and homogeneous, non-laminated dolomite.

CENTIMETRES
16.35.115.2W5 II
1 4
a

1

a
16.35.115.2W5 CORE II BOXES 1 4 ↑ 1 3

2

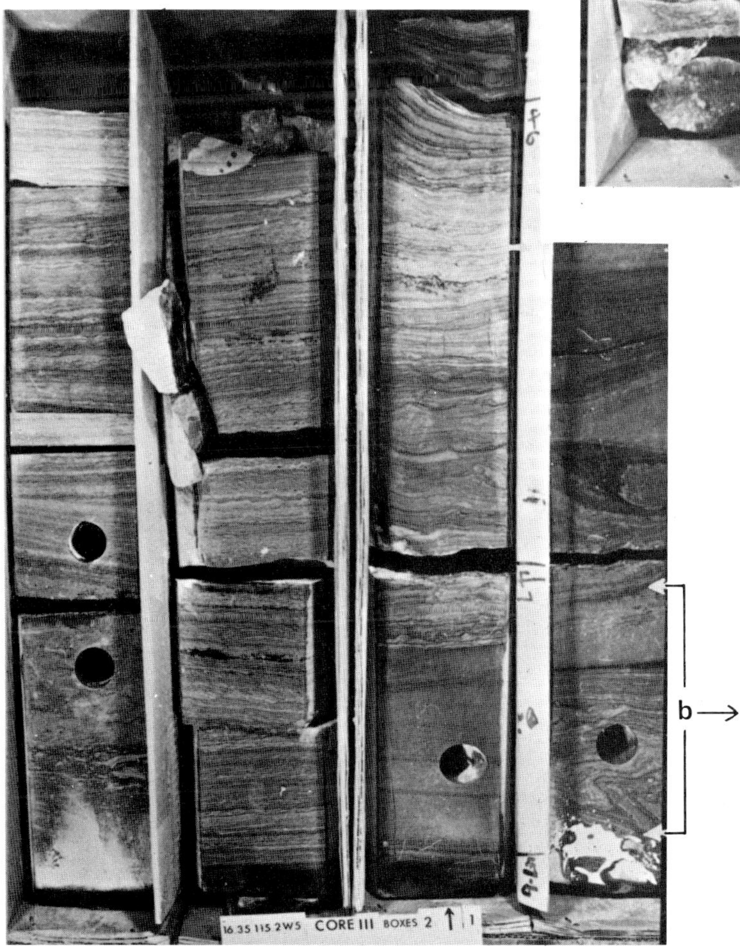
46
47
16.35.115.2W5 CORE III BOXES 2 ↑ 1
b →

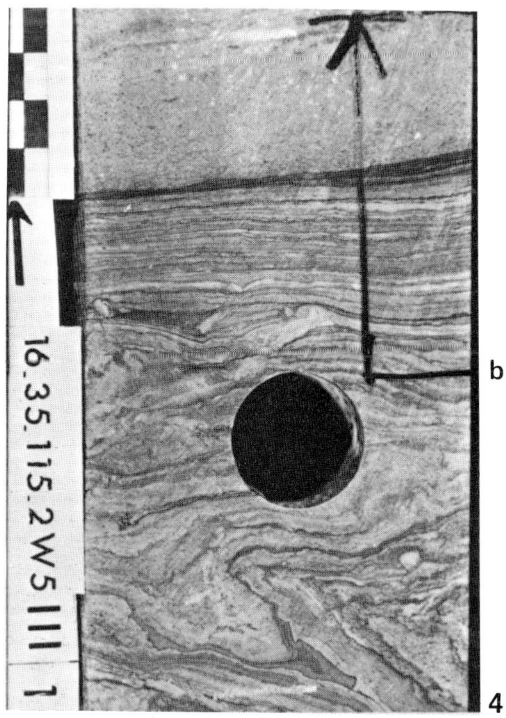
16.35.115.2W5 III
1
b

4

CHAPITRE V (Planches 31 à 37). *CHAPTER V (Plates 31 to 37)*

Etude séquentielle du toit de la série évaporitique du Lias inférieur dans un puits du bassin Aquitain (Auch 1) France du sud-ouest.

Sequential study of the top of the evaporitic series of the Lower Lias in a well in the Aquitaine Basin (Auch 1), South-Western France.

PLANCHE 31 SECTIONS POLIES (x 1)

Fig. 1 Microconglomérat à galets foncés, se place très localement à la base du faciès 1 (dolomicrite).
Auch 1, C 19, 1 970 m.

Fig. 2 Laminations algaires disloquées par place, en cupules par effet de dessiccation (faciès 1, dolomicrite).
Auch 1, C 19, 1 969,25 m.

Fig. 3 Laminations fines, claires et sombres, presque rectilignes (autre aspect du faciès 1, dolomicrite).
Auch 1, C 18, 1 964,60 m.

Fig. 4 Laminations irrégulières, claires et sombres avec développement de pseudomorphoses de gypse par la calcite (faciès 1 à 2, dolomicrite).
Auch 1, C 19, 1 971,80 m.

PLATE 31 POLISHED SECTIONS (x 1)

Fig. 1 Microconglomerate with dark pebbles, situated very locally at the base of facies 1 (dolomicrite).
Auch 1, C 19, 1 970 m.

Fig. 2 Algae laminations dislocated in place, in "chips" locally due to the effect of dessiccation (facies 1, dolomicrite).
Auch 1, C 19, 1 969,25 m.

Fig. 3 Fine laminations, light and dark, almost rectilinear (another aspect of facies 1, dolomicrite).
Auch 1, C 18, 1 964,60 m.

Fig. 4 Irregular laminations, light and dark, with the development of pseudomorphoses of gypsum by the calcite (facies 1 to 2, dolomicrite).
Auch 1, C 19, 1 971,80 m.

1

2

3

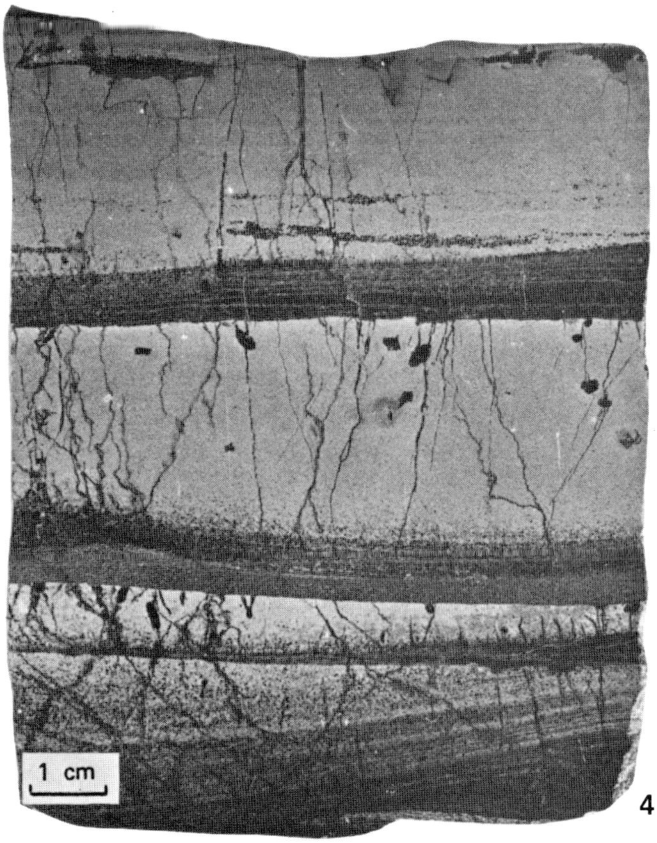

4

CHAPITRE V (Planches 31 à 37). *CHAPTER V (Plates 31 to 37)*

Etude séquentielle du toit de la série évaporitique du Lias inférieur dans un puits du bassin Aquitain (Auch 1) France du sud-ouest.

Sequential study of the top of the evaporitic series of the Lower Lias in a well in the Aquitaine Basin (Auch 1), South-Western France.

PLANCHE 32 LAMES MINCES. FACIES 1. DOLOMICRITES

Fig. 1 Aspect particulier d'un faciès laminé ondulé dont les ondulations localement plus foncées représenteraient un ancien encroûtement de type algaire. En bas quelques craquelures à remplissage de calcite (dessiccation ?).
Vers le haut quelques plages irrégulières calcitiques à rapprocher des *birdseyes* ?
Auch 1, C 17, 1 902,90 m.

Fig. 2 Faciès très irrégulièrement laminé avec microravinement et structure pseudopelletoïdale.
Auch 1, C 21, 2 000 m.

PLATE 32 THIN SECTIONS. FACIES 1. DOLOMICRITES.

Fig. 1 Notable aspect of an ondulated laminated facies, the ondulations which are locally more darkened, would thus represent a former algal type encrusting. At the bottom, some cracks with calcite filling (dessiccation ?). Towards the top, some irregular calcitic patches, which could be compared with birdseyes ?
Auch 1, C 17, 1 902,90 m.

Fig. 2 Very irregularly laminated facies, with microravinement and pseudopelletoïdale structure.
Auch 1, C 21, 2 000 m.

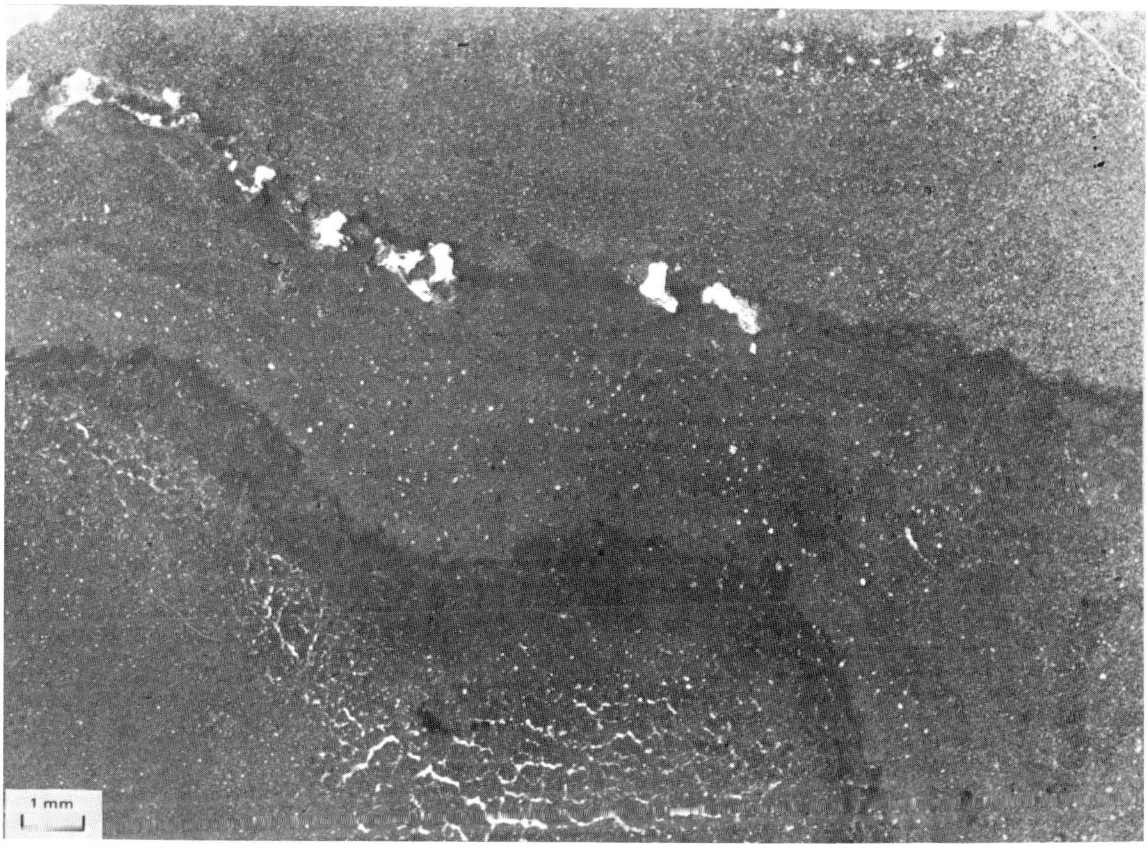

1

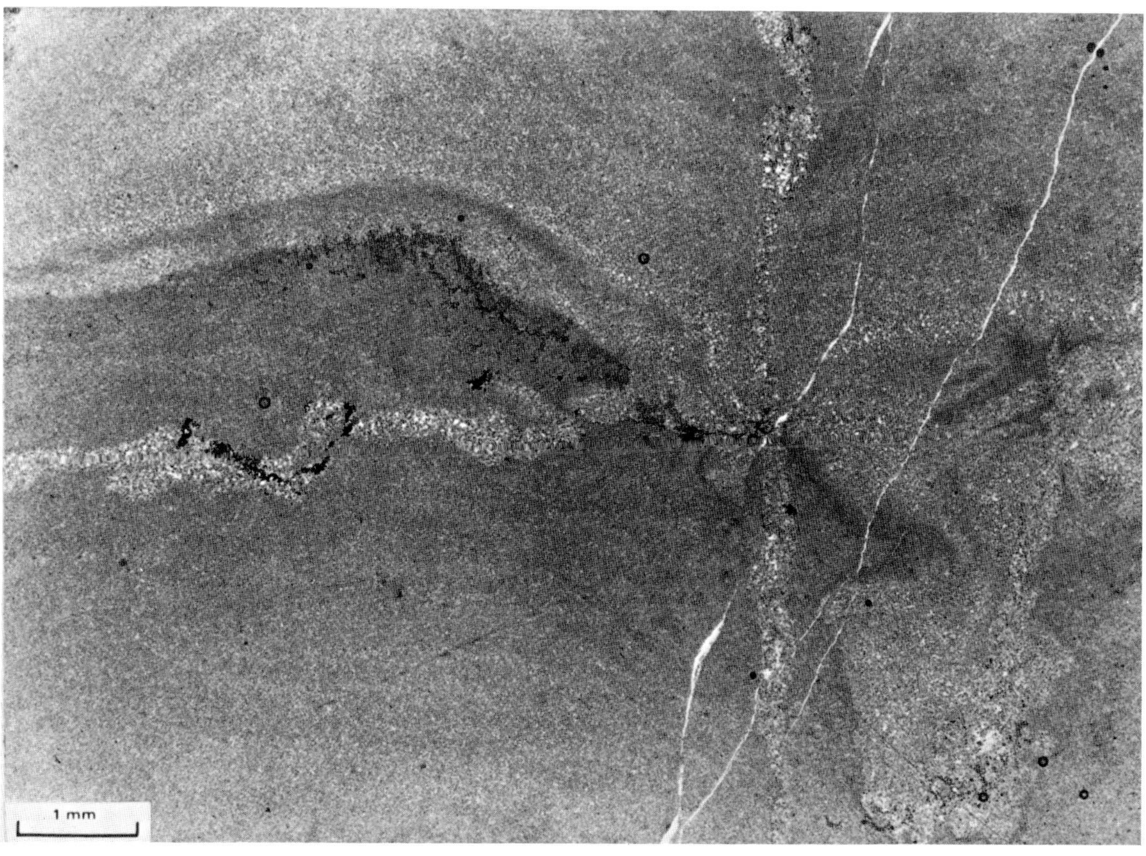

2

CHAPITRE V (Planches 31 à 37). *CHAPTER V (Plates 31 to 37)*

Etude séquentielle du toit de la série évaporitique du Lias inférieur dans un puits du bassin Aquitain (Auch 1) France du sud-ouest.

Sequential study of the top of the evaporitic series of the Lower Lias in a well in the Aquitaine Basin (Auch 1), South-Western France.

PLANCHE 33 SECTION POLIE ET LAMES MINCES ; PSEUDOMORPHOSES DE GYPSE PAR LA CALCITE

Fig. 1 Pseudomorphoses de gypse par la calcite dispersées dans la dolomicrite. Faciès 2 (section polie).
Auch 1, C 19, 1 967,10 m (x 1).

Fig. 2 Aspect microscopique du faciès 2 (plaques minces).
Auch 1, C 19, 1979 m.

Fig. 3 Autre aspect du même faciès entrecoupé de lamines riches en argile et matière organique (plaques minces).
Auch 1, C 19, 1 981 m.

PLATE 33 POLISHED SLAB AND THIN SECTIONS : PSEUDOMORPHS OF GYPSUM BY CALCITE

Fig. 1 Pseudomorphoses of gypsum by calcite, which are dispersed in the dolomicrite. Facies 2 (polished section).
Auch 1, C 19, 1 967, 10 m (x 1).

Fig. 2 Microscopical aspect of facies 2 (thin sections).
Auch 1, C 19, 1 979 m.

Fig. 3 Another aspect of the same facies intersected by lamines rich in clay and in organic matter (thin sections).
Auch 1, C 19, 1 981 m.

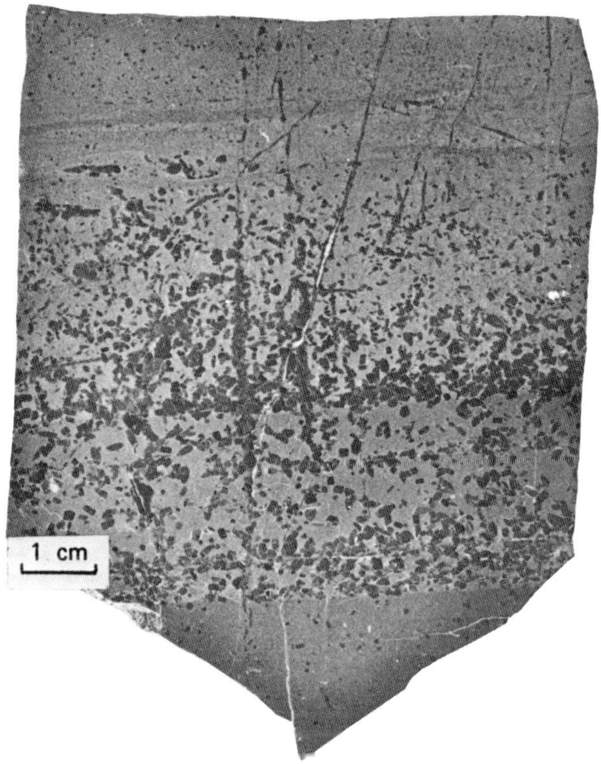

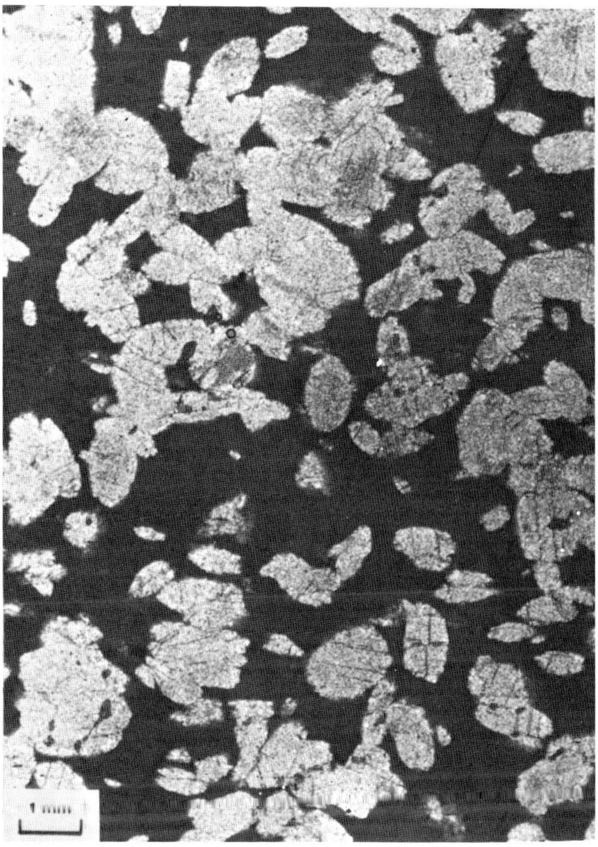

2

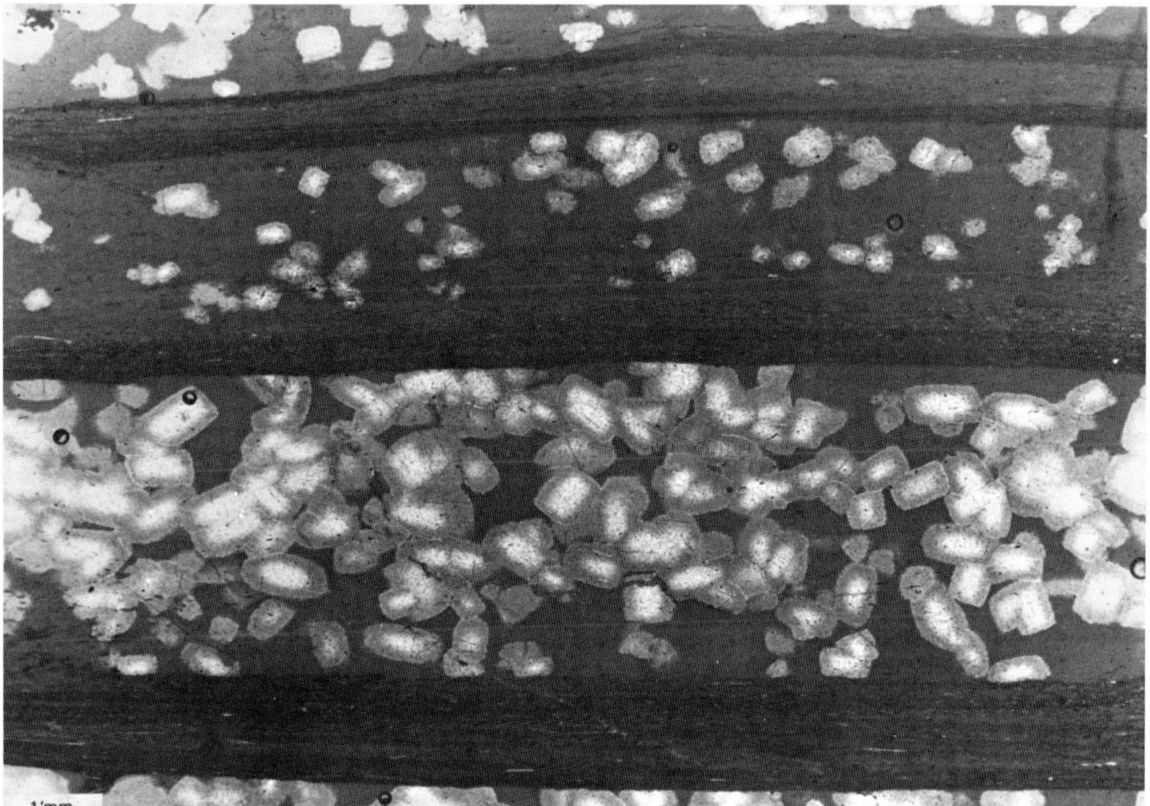

3

CHAPITRE V (Planches 31 à 37). *CHAPTER V (Plates 31 to 37)*

Etude séquentielle du toit de la série évaporitique du Lias inférieur dans un puits du bassin Aquitain (Auch 1) France du sud-ouest.

Sequential study of the top of the evaporitic series of the Lower Lias in a well in the Aquitaine Basin (Auch 1), South-Western France.

PLANCHE 34 PLAQUES MINCES. FACIES 3. DOLOMICRITES A NODULES D'ANHYDRITE

Fig. 1 Nodules macrocristallins d'anhydrite dans la dolomicrite.
Auch 1, C 22, 2 031,30 m.

Fig. 2 Nodules microcristallins et à contours diffus d'anhydrite dans la dolomicrite.
Auch 1, C 20, 1 985,60 m.

PLATE 34 THIN SECTIONS. FACIES 3. DOLOMICRITE WITH ANHYDRITE NODULES

Fig. 1 Macrocrystalline nodules of anhydrite in the dolomicrite.
Auch 1, C 22, 2 031,30 m.

Fig. 2 Microcrystalline nodules with diffused contours of anhydrite in the dolomicrite.
Auch 1, C 20, 1 985,60 m.

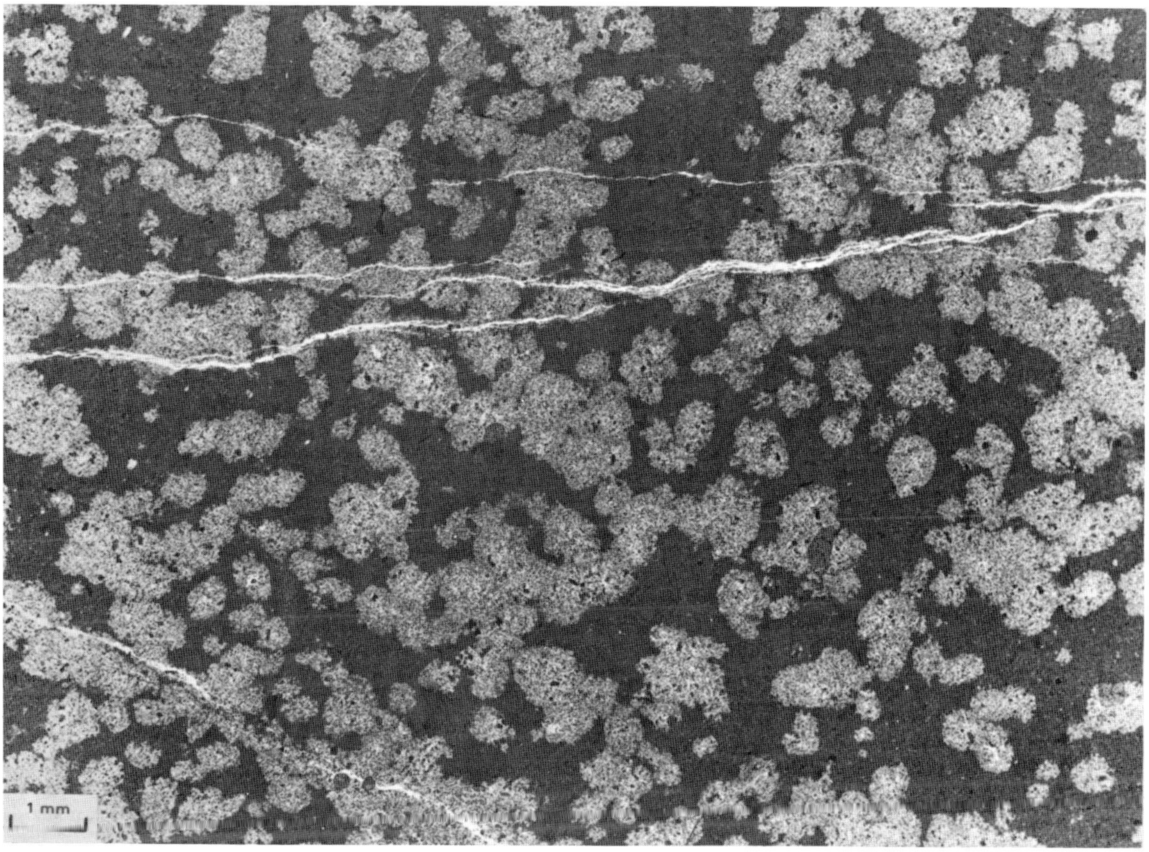

1

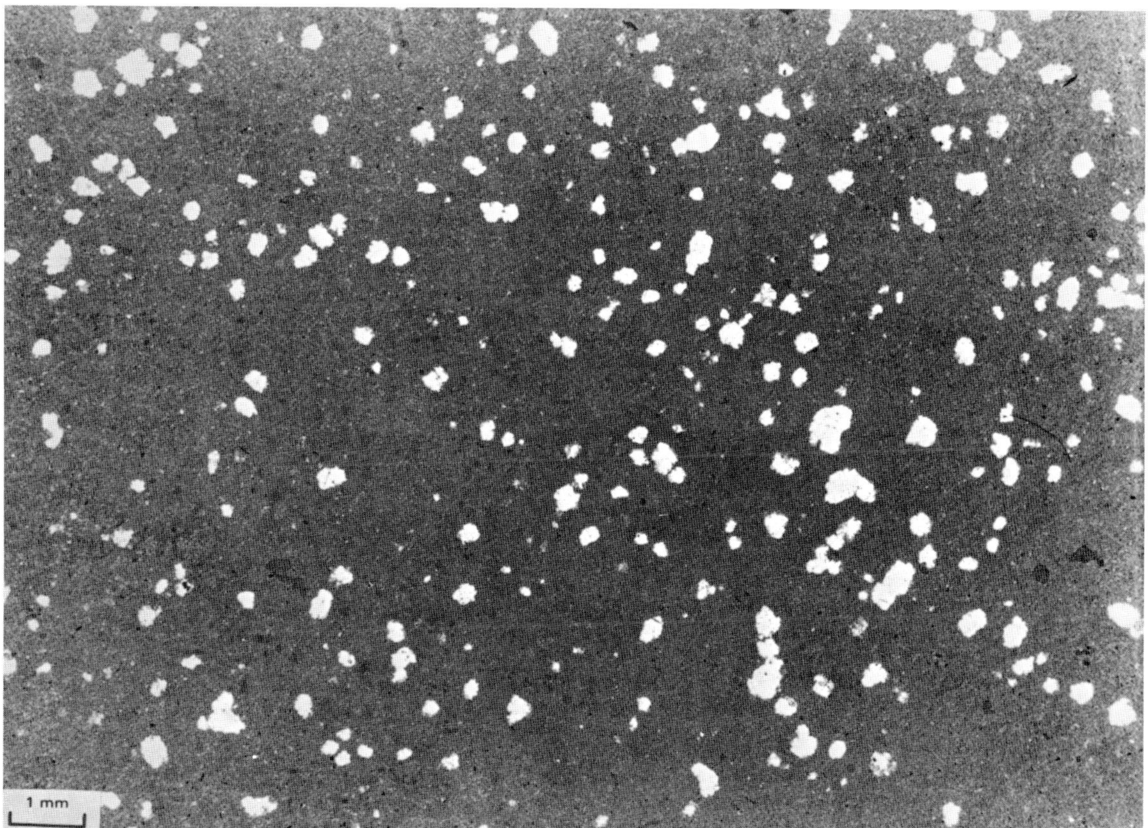

2

CHAPITRE V (Planches 31 à 37). *CHAPTER V (Plates 31 to 37)*

Etude séquentielle du toit de la série évaporitique du Lias inférieur dans un puits du bassin Aquitain (Auch 1) France du sud-ouest.

Sequential study of the top of the evaporitic series of the Lower Lias in a well in the Aquitaine Basin (Auch 1), South-Western France.

PLANCHE 35 SECTIONS POLIES. FACIES 5 (× 1). ANHYDRITE A STRUCTURE EN GLOMERULES ASSOCIES, PUIS CONTOURNEE OU ENTEROLITHIQUE

Fig. 1 Auch 1, C 18, 1 965,80 m.

Fig. 2 Auch 1, C 22, 2 027,30 m.

PLATE 35 POLISHED SLABS. FACIES 5 (× 1). ANHYDRITE WITH A STRUCTURE OF ASSOCIATED GLOMERULES, WHICH ARE THEN CONTORTED OR ENTEROLITHIC

Fig. 1 Auch 2, C 18, 1 965,80 m.

Fig. 2 Auch 1, C 22, 2 027,30 m.

1

2

CHAPITRE V (Planches 31 à 37). *CHAPTER V (Plates 31 to 37)*

Etude séquentielle du toit de la série évaporitique du Lias inférieur dans un puits du bassin Aquitain (Auch 1) France du sud-ouest.

Sequential study of the top of the evaporitic series of the Lower Lias in a well in the Aquitaine Basin (Auch 1), South-Western France.

PLANCHE 36 SECTION POLIE, FACIES 6 (× 1). ANHYDRITE A STRUCTURE EN RESEAU OU "CHICKEN-WIRE"

Auch 1, C 22, 2 026,85 m.

PLATE 36 POLISHED SLAB, FACIES 6 (× 1). ANHYDRITE WITH A NETWORK OR CHICKEN-WIRE STRUCTURE

Auch 1, C 22, 2 026,85 m.

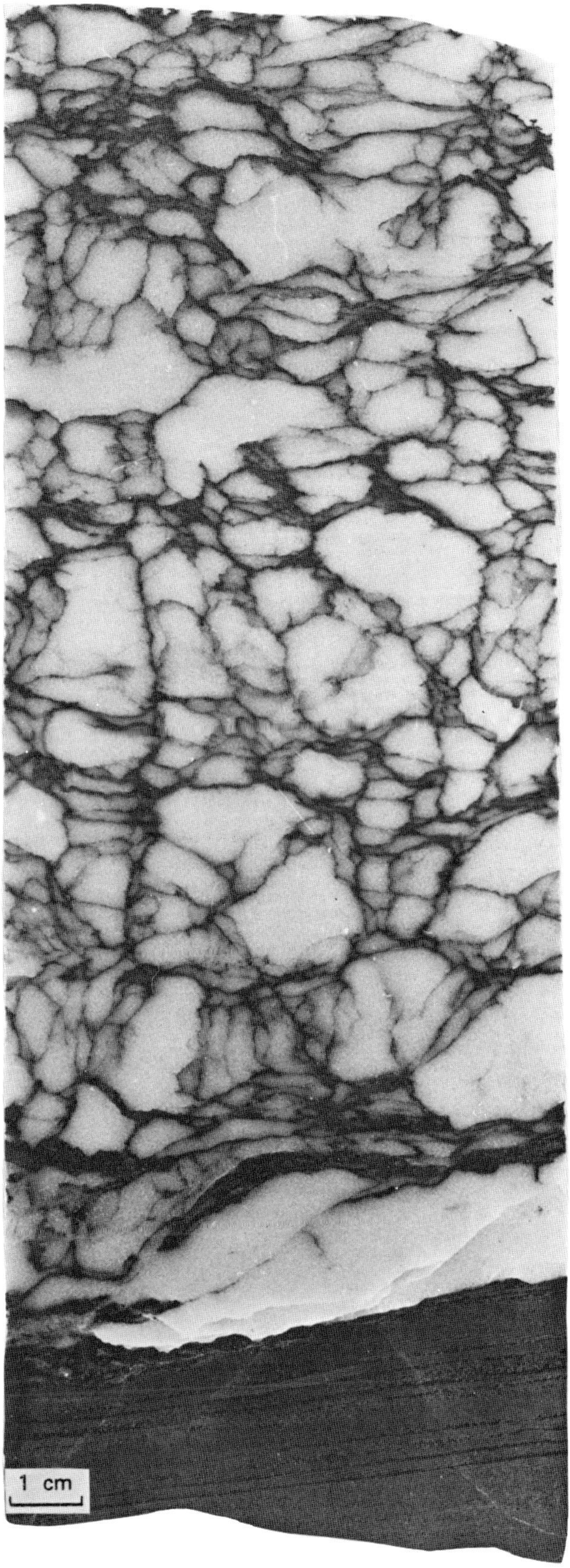
1 cm

CHAPITRE V (Planches 31 à 37). *CHAPTER V (Plates 31 to 37)*

Etudc séquentielle du toit de la série évaporitique du Lias inférieur dans un puits du bassin Aquitain (Auch 1) France du sud-ouest.

Sequential study of the top of the evaporitic series of the Lower Lias in a well in the Aquitaine Basin (Auch 1), South-Western France.

PLANCHE 37 PLAQUES MINCES, FACIES 6, EN "CHICKEN-WIRE"

Aspects microscopiques de la structure en *chicken-wire* avec anhydrite en aiguilles ou bâtonnets assemblés en glomérules plus ou moins jointifs.

Fig. 1 Auch 1, C 22, 2 022 m.

Fig. 2 Auch 2, C 22, 2 045,5 m.

PLATE 37 THIN SECTIONS, FACIES 6, CHICKEN-WIRE FACIES

Microscopical aspects of the chicken-wire structure with anhydrite in needles or lath, assembled in more or less joined glomerules (nodular-mosaic).

Fig. 1 Auch 1, C 22, 2 022 m.

Fig. 2 Auch 1, C 22, 2 045,5 m.

1

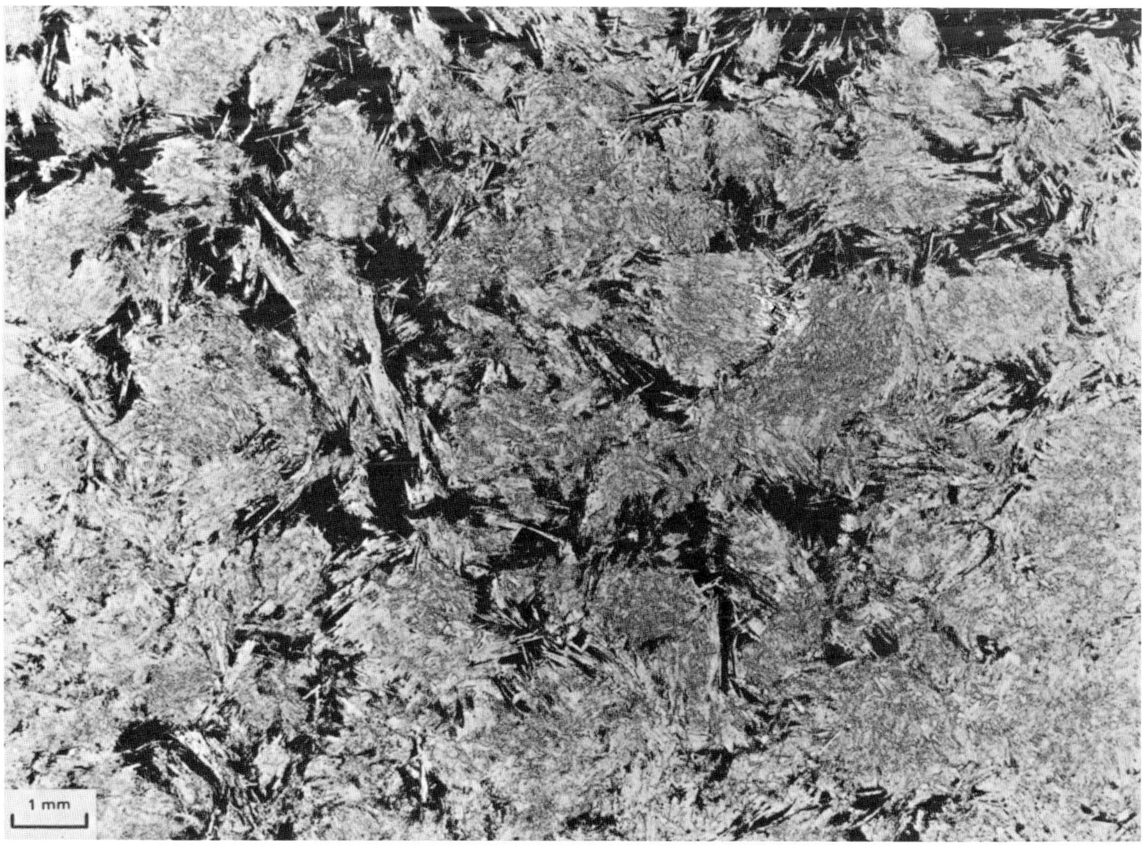

2

CHAPITRE VI (Planches 38 à 44). *CHAPTER VI (Plates 38 to 44)*

Les séquences évaporitiques du Miocène terminal de Sicile et d'Espagne méridionale.

The evaporitic sequences of the terminal Miocene of Sicily and of Southern Spain.

PLANCHE 38 LES SEQUENCES DE LA SERIE GYPSEUSE SUPERIEURE EN SICILE

Fig. 1 Illustration de l'une des séquences de la série supérieure choisie sur les marges du bassin (Palma di Montecchiaro) ; on peut observer de droite à gauche sur la photo : le sommet du banc de gypse sélénite de la séquence sous-jacente (Gi), des argiles et des marnes associées à des niveaux tripolacés (Mt), du gypse finement lité (balatino : Gb) et le banc de gypse en chevrons (sélénite : Gs). La base du banc supérieur de gypse sélénite présente une limite en dents de scie dont les parties aiguës correspondent à la base des cristaux de sélénite groupés en éventail divergent vers le haut.

Fig. 2 Détail d'un cristal de gypse sélénite (Montedoro).

Fig. 3 et 4 Détail de gypse finement lité (ou balatino). Haut : litage horizontal plan-parallèle (Montedoro). Bas : litage ondulé (Eraclea Minoa).

PLATE 38 SEQUENCES OF THE UPPER GYPSIFEROUS SERIES IN SICILY

Fig. 1 Illustrations of one of the sequences of the upper series, chosen on the edges of the basin (Palma di Montecchiaro) ; from right to left on the photo can be noted : the top of the bank of selenite gypsum of the underlying sequence (Gi), clays and marls associated with the tripolaced levels (Mt), finely bedded gypsum (balatino : Gb) and the bank of gypsum-chevrons (selenite : Gs). The base of the upper bank of selenite gypsum presents a "saw-teeth" boundary, of which the pointed parts correspond to the base of the selenite crystals grouped in a pan, diverging towards the top.

Fig. 2 Detail of a selenite crystal (Montedoro).

Figs. 3 and 4 Detail of finely bedded gypsum (or balatino) top : horizontal bedding, parallel plane (Montedoro). Bottom : undulated bedding (Eraclea Minoa).

1

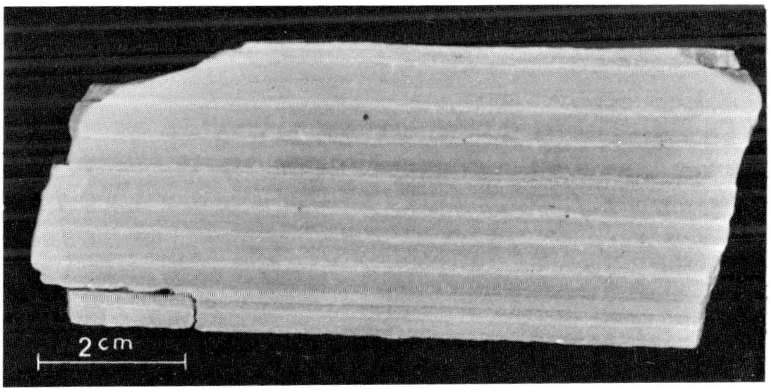

3

4

CHAPITRE VI (Planches 38 à 44). *CHAPTER VI (Plates 38 to 44)*

Les séquences évaporitiques du Miocène terminal de Sicile et d'Espagne méridionale.

The evaporitic sequences of the terminal Miocene of Sicily and of Southern Spain.

PLANCHE 39 STRUCTURES SEDIMENTAIRES DU GYPSE BALATINO ET GYPSES ASSOCIES

Différents types de structures sédimentaires affectant les gypses finement lités (balatino) et les autres types de gypse associés (gypsarénite et gypse microcristallin) dans les séquences de la "série supérieure".

Fig. 1 *Ripple-marks* dans le gypse lité (Palma di Montecchiaro).

Fig. 2 Structures de glissement déversées vers la partie centrale du bassin, affectant les gypses finement lités entre deux bancs horizontaux de gypsarénite (Santa Elisabeta).

Fig. 3 Phénomène de *slumping* dans un banc de gypse microcristallin associé au gypse finement lité (San Cataldo).

Fig. 4 Remaniements *in-situ* de gypse finement lité à la base d'un banc de gypsarénite avec stries verticales de dissolution actuelles (Siculiana).

Fig. 5 Stratifications obliques dans un banc de gypse détritique interstratifié entre deux niveaux de gypse finement lité (près de Montedoro).

PLATE 39 SEDIMENTARY STRUCTURES OF BALATINO GYPSUM AND ASSOCIATED GYPSUM

Different types of sedimentary structures affecting the finely bedded gypsum (balatino) and the other types of associated gypsum (gypsarenite and microcrystalline gypsum) in the sequences of the "upper serie".

Fig. 1 Ripple-marks in the bedded gypsum (Palma di Montecchiaro).

Fig. 2 Structures of sliding inclined towards the central part of the basin, affecting the finely gypsum between two horizontal banks of gypsarenite (Santa Elisabeta).

Fig. 3 Phenomenon of slumping in a bank of microcrystalline gypsum associated with finely bedded gypsum (San Cataldo).

Fig. 4 Reworkings *in-situ* of finely bedded gypsum at the base of a bank of gypsarenite with vertical scratches of present-day dissolution (Siculiana).

Fig. 5 Oblique stratifications in a bank of detritic gypsum unterstratified between the layer of finely bedded gypsum (near Montedoro).

1

2

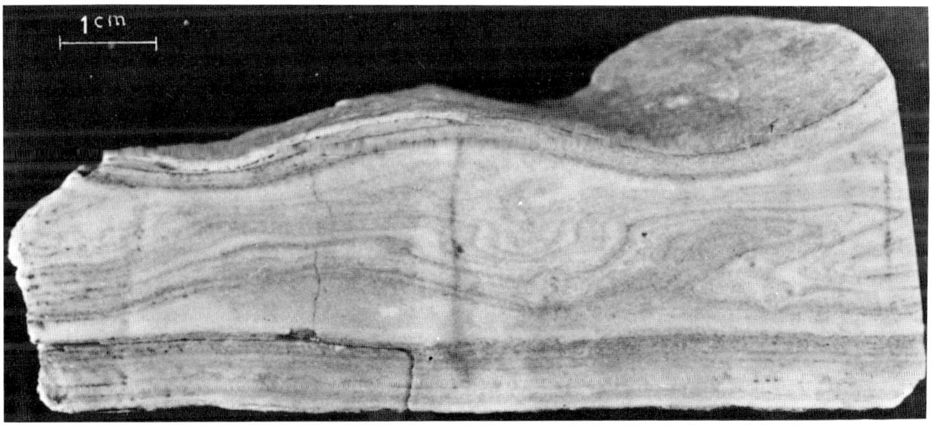

3

4

5

PLANCHE 40 STRUCTURES DU GYPSE SELENITE

Structures particulières du gypse sélénite (Monte Palco au sud de San Cataldo).

Fig. 1 Groupement en éventail des cristaux de sélénite.

Fig. 2 Surface d'un banc de gypse sélénite montrant le sommet des structures en coupoles dues au type de cristallisation illustré par la figure précédente.

Fig. 3 Détail d'une coupole montrant un réseau de fentes radiales de tension dues au "gonflement" tardif de ces structures.

PLATE 40 SELENITE GYPSUM STRUCTURES

Special structures of the selenite gypsum (Monte Palco, to the south of San Cataldo).

Fig. 1 Grouping of the selenite crystals in a fan.

Fig. 2 Surface of a bank of selenite gypsum, showing the top of the dome structures, due to the type of crystallization illustrated in the preceding figure.

Fig. 3 Detail of a dome, showing a network of radial slits of tension, due to the late "swelling" of these structures.

1

2

3

CHAPITRE VI (Planches 38 à 44). *CHAPTER VI (Plates 38 to 44)*

Les séquences évaporitiques du Miocène terminal de Sicile et d'Espagne méridionale.

The evaporitic sequences of the terminal Miocene of Sicily and of Southern Spain.

PLANCHE 41 LES GYPSES DE TYPE NODULAIRE EN SICILE

Fig. 1 Nodules gypseux dans une matrice carbonatée (Santa Caterina Villarmosa).

Fig. 2 Gypse à texture en "mosaïque", lité et contourné (Santa Caterina Villarmosa).

Fig. 3 Gypse à texture floconneuse et en "mosaïque" massive (Eraclea Minoa).

Fig. 4 Nodules gypseux blancs dans une matrice également gypseuse (Casteltermini) évoquant par leur forme certains *birds eyes* de milieu carbonaté.

PLATE 41 THE NODULAR GYPSUMS IN SICILY

Fig. 1 Gypsum nodules in a carbonate matrix (Santa Caterina Villarmosa).

Fig. 2 Gypsum with a "mosaic" texture, bedded and contorted (Santa Caterina Villarmosa).

Fig. 3 Gypsum with a flaky texture, in a large "mosaic" (Eraclea Minoa).

Fig. 4 White gypsum nodules in a similarly gypsum matrix (Casteltermini) recalling certain "birds-eyes" of the carbonate environment, by their form.

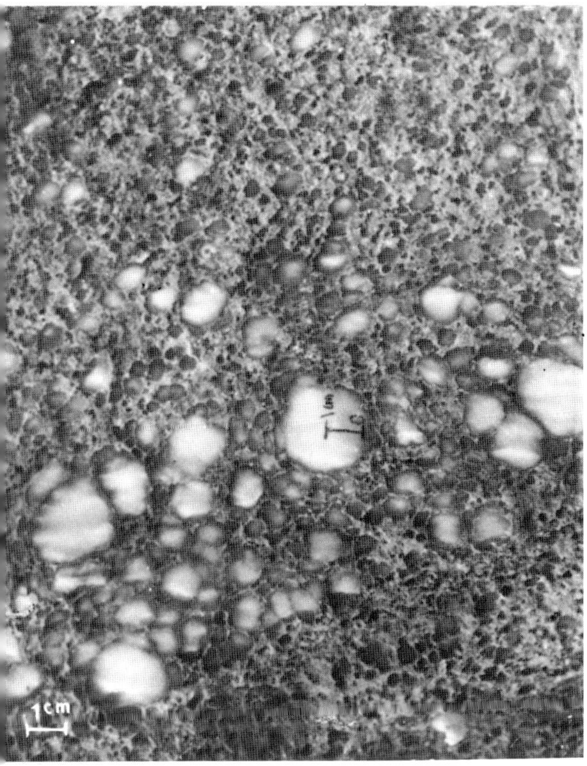

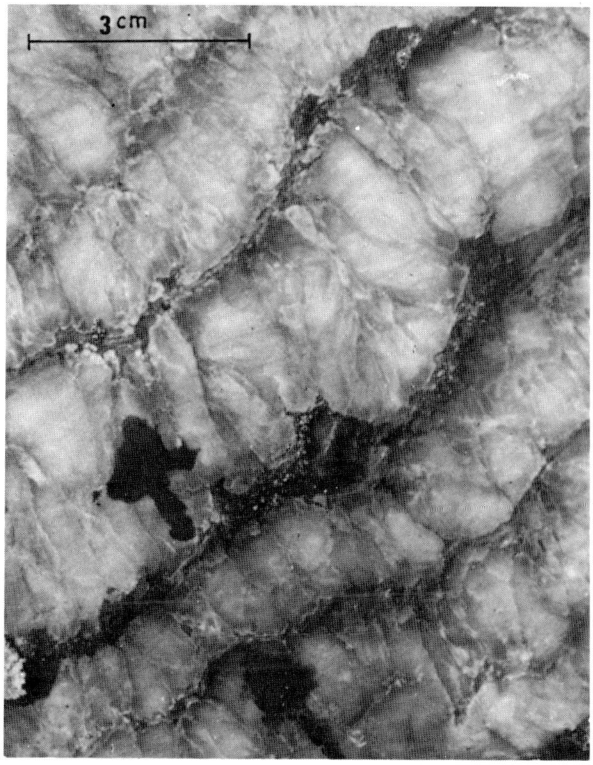

2

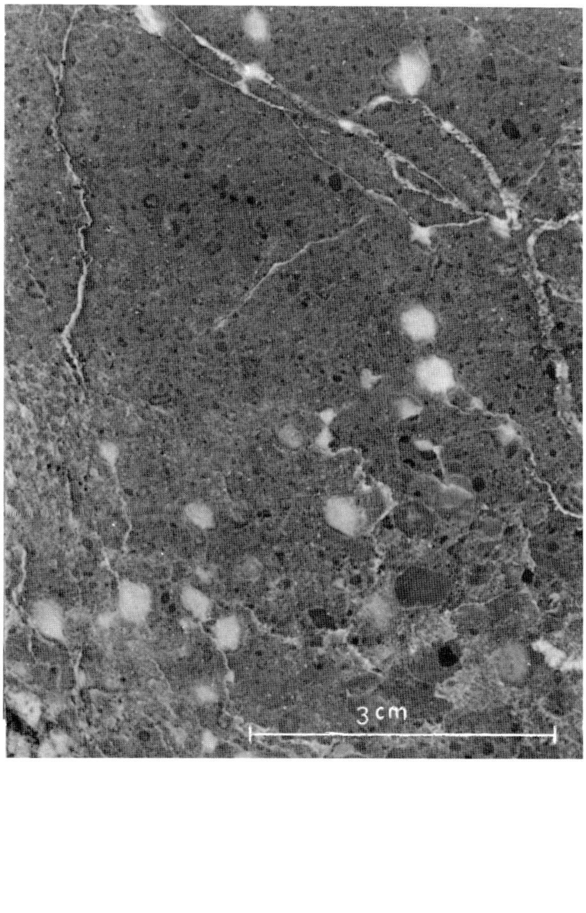

4

CHAPITRE VI (Planches 38 à 44). *CHAPTER VI (Plates 38 to 44)*

Les séquences évaporitiques du Miocène terminal de Sicile et d'Espagne méridionale.

The evaporitic sequences of the terminal Miocene of Sicily and of Southern Spain.

PLANCHE 42 CARACTERES PETROGRAPHIQUES DE LA SERIE GYPSEUSE INFERIEURE

Illustration de quelques caractères pétrographiques de la série inférieure en Sicile, à différentes échelles d'observation.

Fig. 1 Alternance de lits décimétriques de gypse sélénite et de passées carbonatées (généralement dolomitiques), azoïques à grain fin (région de Sutera).

Fig. 2 Passage vertical de gypse sélénite bien cristallisé à un carbonate azoïque, parfois finement laminé. Il s'agit là d'une illustration à l'échelle de l'échantillon d'un type de séquence répétitive très développée dans la série inférieure où elles atteignent plusieurs dizaines de mètres (Sutera).

Fig. 3 Développement diagénétique précoce de cristaux de gypse dans de fines alternances gypse-carbonate (Montallegro).

Fig. 4 Relations gypse-carbonate à l'échelle du microscope électronique dans le même échantillon. Les cristaux de gypse (G) englobent les grains carbonatés (C) dont l'empreinte reste imprimée dans le cristal (Montallegro).

PLATE 42 PETROGRAPHIC FEATURES OF THE LOWER GYPSIFEROUS SERIE

Illustration of some petrographical features of the lower serie in Sicily, at different scales of observation.

Fig. 1 Alternation between decimetric beds of selenite gypsum and of carbonate layers (generally dolomitic), azoic with fine grain (Sutera region).

Fig. 2 Vertical passage of well crystallized gypsum with azoic carbonate, sometimes finely laminated. This is an illustration by sample observation, of a type of repetitive sequence, very developed in the lower serie, where they attain several tens of meters (Sutera).

Fig. 3 Early diagenetic development of gypsum crystals in fine gypsum-carbonate alternations.

Fig. 4 Gypsum-carbonate relations viewed by electronic microscope in the same sample. The gypsum crystals (G) englobe the carbonate grains (C), the empreint of which remains in the crystal (Montallegro).

1

2

3

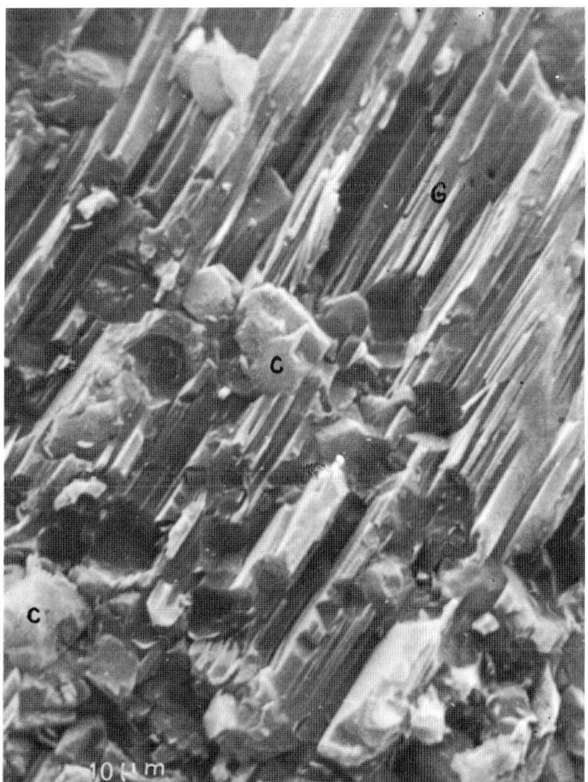

4

CHAPITRE VI (Planches 38 à 44). *CHAPTER VI (Plates 38 to 44)*

Les séquences évaporitiques du Miocène terminal de Sicile et d'Espagne méridionale.

The evaporitic sequences of the terminal Miocene of Sicily and of Southern Spain.

PLANCHE 43 STRUCTURES DES GYPSES DES BASSINS INTERNES D'ESPAGNE MERIDIONALE

Structures des gypses dans les bassins internes d'Espagne méridionale (exemple du bassin de Murcie, San Miguel de Salinas et Benejuzar).

Fig. 1 Gypse de type sélénite en très grands cristaux.

Fig. 2 Gypse blanc à texture en "mosaïque" se développant aux dépens du gypse sélénite.

Fig. 3 Relations sélénite-nodulaire à l'échelle de l'échantillon.

Fig. 4 Relations du gypse sélénite avec les marnes sus-jacentes.

PLATE 43 GYPSUM STRUCTURES IN THE INTERNAL BASINS OF SOUTHERN SPAIN

Gypsum structures in the internal basins of Southern Spain (example of the Murcie, San Miguel de Salinas and Benejuzar basin).

Fig. 1 Gypsum of selenite type in very large crystals.

Fig. 2 White gypsum of "mosaic" texture, developing into the selenite gypsum.

Fig. 3 Selenite-nodular relations on the sample.

Fig. 4 Relation between the selenite gypsum and the overlying marls.

1

2

3

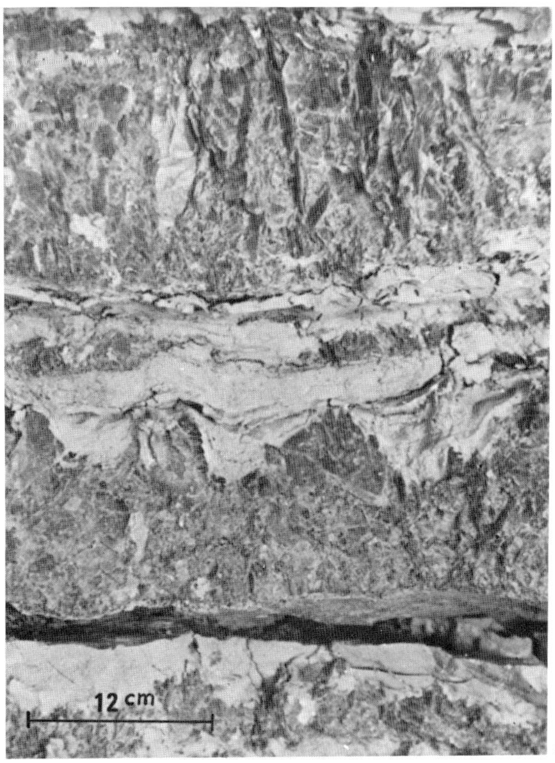

4

CHAPITRE VI (Planches 38 à 44). *CHAPTER VI (Plates 38 to 44)*

Les séquences évaporitiques du Miocène terminal de Sicile et d'Espagne méridionale.

The evaporitic sequences of the terminal Miocene of Sicily and of Southern Spain.

PLANCHE 44 STRUCTURES DE GYPSES DES BASSINS EXTERNES D'ESPAGNE MERIDIONALE

Structures particulières de gypses dans les bassins externes d'Espagne méridionale.

Fig. 1 Structures évoquant des figures de courant de type *flute cast* à la base de niveaux de gypse finement lité à laminations carbonatées (Lorca).

Fig. 2 Gypse lité nodulaire (bassin de Grenade, La Mala).

Fig. 3 Structure entérolithique dans le gypse nodulaire du bassin de Grenade (près de Cacin).

PLATE 44 GYPSUM STRUCTURES IN THE EXTERNAL BASINS OF SOUTHERN SPAIN

Notable gypsum structures in the external basins of Southern Spain.

Fig. 1 Structures recalling the current figures of the "flute-cast" type, at the base of the finely bedded gypsum layer with carbonate laminations (Lorca).

Fig. 2 Nodular bedded gypsum (Grenada Basin, La Mala).

Fig. 3 Enterolithic structure in the nodular gypsum of the Grenada Basin (near Cacin).

1

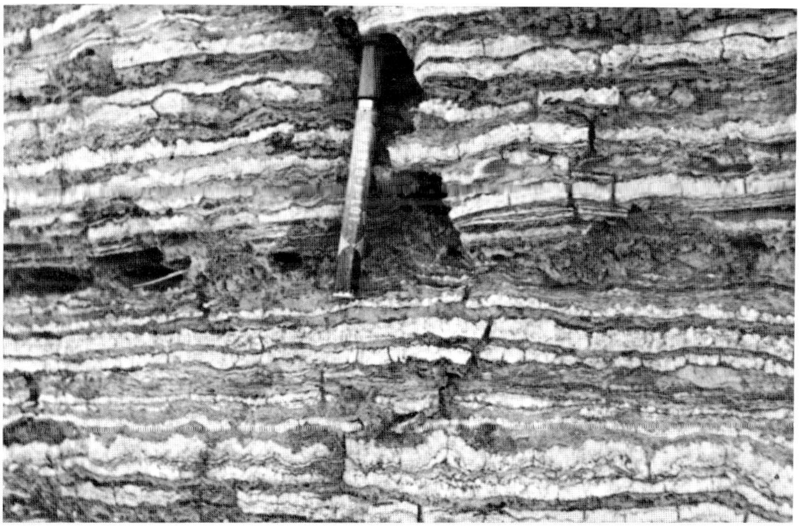

2

3

CHAPITRE VII (Planches 45 à 64). *CHAPTER VII (Plates 45 to 64)*

Eléments de pétrographie des évaporites oligocènes des bassins de la Bresse et de Valence (Est de la France, vallée de la Saône et du Rhône).

Petrographical data of Oligocene evaporites in the Bresse and Valence Basins.

PLANCHE 45 EXEMPLES D'ALTERNANCES INSOLUBLES-SEL (VALENCE)

Fig. 1 et 2 Alternance de lits horizontaux d'argiles et de sel renfermant quelques nodules d'anhydrite (Valence).

PLATE 45 EXAMPLES OF ALTERNATION INSOLUBLES-SALT (VALENCE)

Figs. 1 and 2 Alternation of horizontal layers of clay and salt, enclosing some anhydrite nodules (Valence).

1

2

CHAPITRE VII (Planches 45 à 64). *CHAPTER VII (Plates 45 to 64)*

Eléments de pétrographie des évaporites oligocènes des bassins de la Bresse et de Valence (Est de la France, vallée de la Saône et du Rhône).

Petrographical data of Oligocene evaporites in the Bresse and Valence Basins.

PLANCHE 46 EXEMPLES D'ALTERNANCES INSOLUBLES-SEL (BRESSE)

La base de l'échantillon est représentée par une alternance de lits argileux horizontaux et de sel microcristallin de teinte grisâtre.

Vers le sommet le sel devient plus pur et est formé de gros cristaux ; les lits horizontaux disparaissent (Bresse).

PLATE 46 EXAMPLES OF ALTERNATION INSOLUBLES-SALT (VALENCE)

The base of the sample is made up of horizontal clay and greyish microcrystalline salt.

Towards the top, the salt becomes purer and composed of large crystals ; the horizontal layers disappear (Bresse).

1 cm

CHAPITRE VII (Planches 45 à 64). *CHAPTER VII (Plates 45 to 64)*

Eléments de pétrographie des évaporites oligocènes des bassins de la Bresse et de Valence (Est de la France, vallée de la Saône et du Rhône).

Petrographical data of Oligocene evaporites in the Bresse and Valence Basins.

PLANCHE 47 SEL DE RECRISTALLISATION (BRESSE)

Sel de recristallisation : cristaux de sel gemme imbriqués les uns dans les autres et séparés par des plages argileuses ou anhydritiques (Bresse).

PLATE 47 RECRYSTALLISATION SALT (BRESSE)

Imbricated salt crystals, separated by clay or anhydrite patches (Bresse).

1 cm

CHAPITRE VII (Planches 45 à 64). *CHAPTER VII (Plates 45 to 64)*

Eléments de pétrographie des évaporites oligocènes des bassins de la Bresse et de Valence (Est de la France, vallée de la Saône et du Rhône).

Petrographical data of Oligocene evaporites in the Bresse and Valence Basins.

PLANCHE 48 HALITE DE RECRISTALLISATION AU MICROSCOPE OPTIQUE ET A BALAYAGE

Fig. 1 Halite de recristallisation (Bresse) : cristal de sel gemme bordé de baguettes d'anhydrite associées à des rhomboèdres de calcite.

Fig. 2 Cristaux d'anhydrite allongés inclus dans le sel.

Fig. 3 Cristaux aciculaires d'anhydrite au microscope électronique à balayage.

Fig. 4 Contact entre un cristal d'halite et des aiguilles d'anhydrite.

PLATE 48 RECRYSTALLISATION HALITE THROUGH OPTICAL AND SCANNING ELECTRON MICROSCOPE

Fig. 1 Recrystallisation halite (Bresse) : salt crystal surrounded by anhydrite laths associated with calcite rhomboedrons.

Fig. 2 Lengthened anhydrite crystals included in the salt.

Fig. 3 Acicular anhydrite crystals through scanning electron microscope.

Fig. 4 Contact between a halite crystal and anhydrite needles.

1

x 59,5

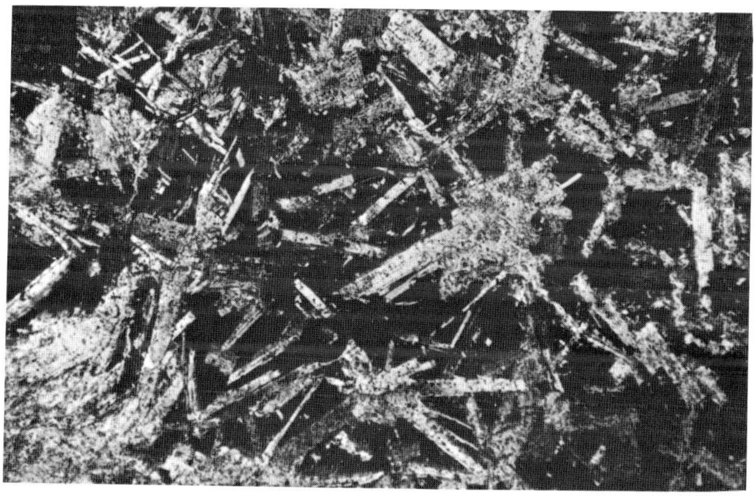

2

x 51

3

x 4 250

4

x 1700

CHAPITRE VII (Planches 45 à 64). *CHAPTER VII (Plates 45 to 64)*

Eléments de pétrographie des évaporites oligocènes des bassins de la Bresse et de Valence (Est de la France, vallée de la Saône et du Rhône).

Petrographical data of Oligocene evaporites in the Bresse and Valence Basins.

PLANCHE 49 ANHYDRITE NODULAIRE

Fig. 1 Anhydrite à structure *chicken wire.*

Fig. 2 Structure gneissique d'un nodule d'anhydrite.

PLATE 49 NODULAR ANHYDRITE

Fig. 1 Anhydrite with "chicken wire" structure.

Fig. 2 Gneissic structure of an anhydrite nodule.

1

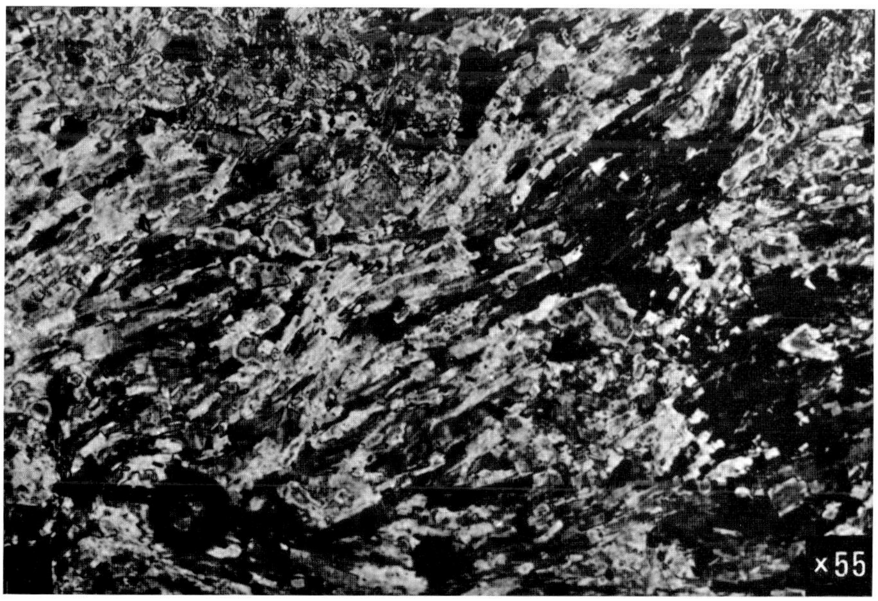

2

CHAPITRE VII (Planches 45 à 64). *CHAPTER VII (Plates 45 to 64)*

Eléments de pétrographie des évaporites oligocènes des bassins de la Bresse et de Valence (Est de la France, vallée de la Saône et du Rhône).

Petrographical data of Oligocene evaporites in the Bresse and Valence Basins.

PLANCHE 50 FORMATION D'UN NODULE PAR DESAGREGATION D'UNE ANHYDRITE EN MOSAIQUE

PLATE 50 NODULE FORMATION BY DISINTEGRATION OF MOSAIC ANHYDRITE

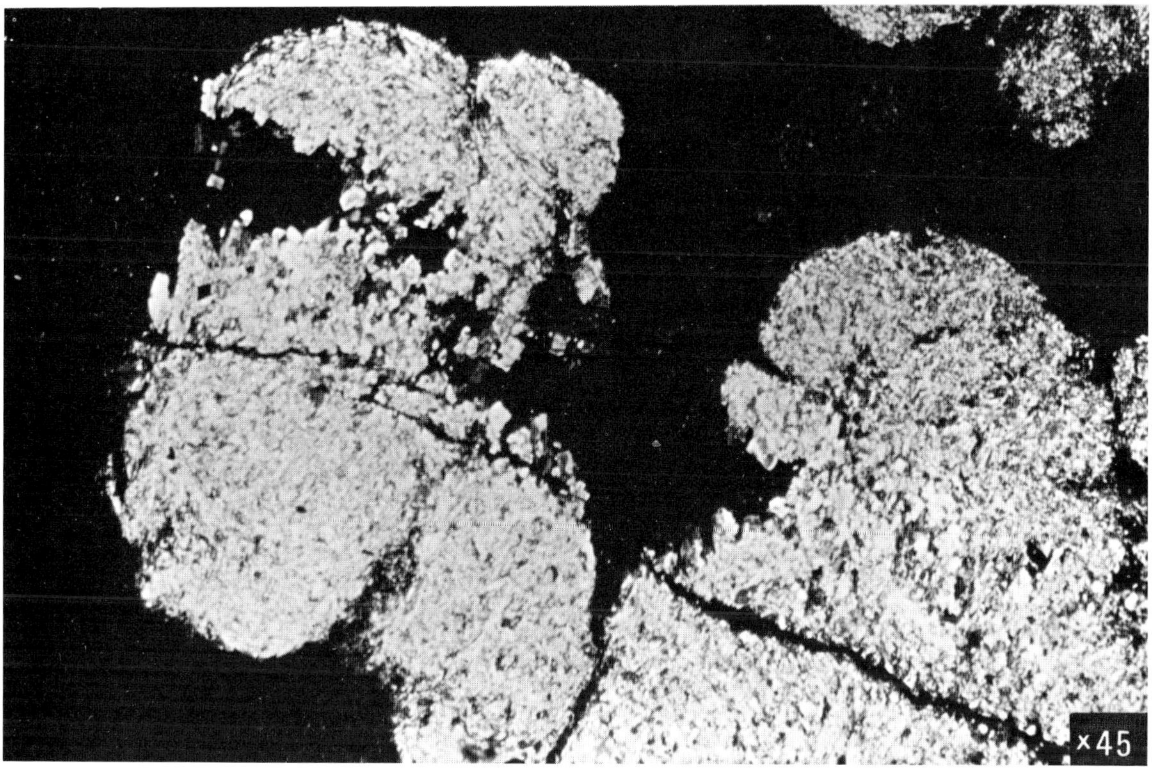

1

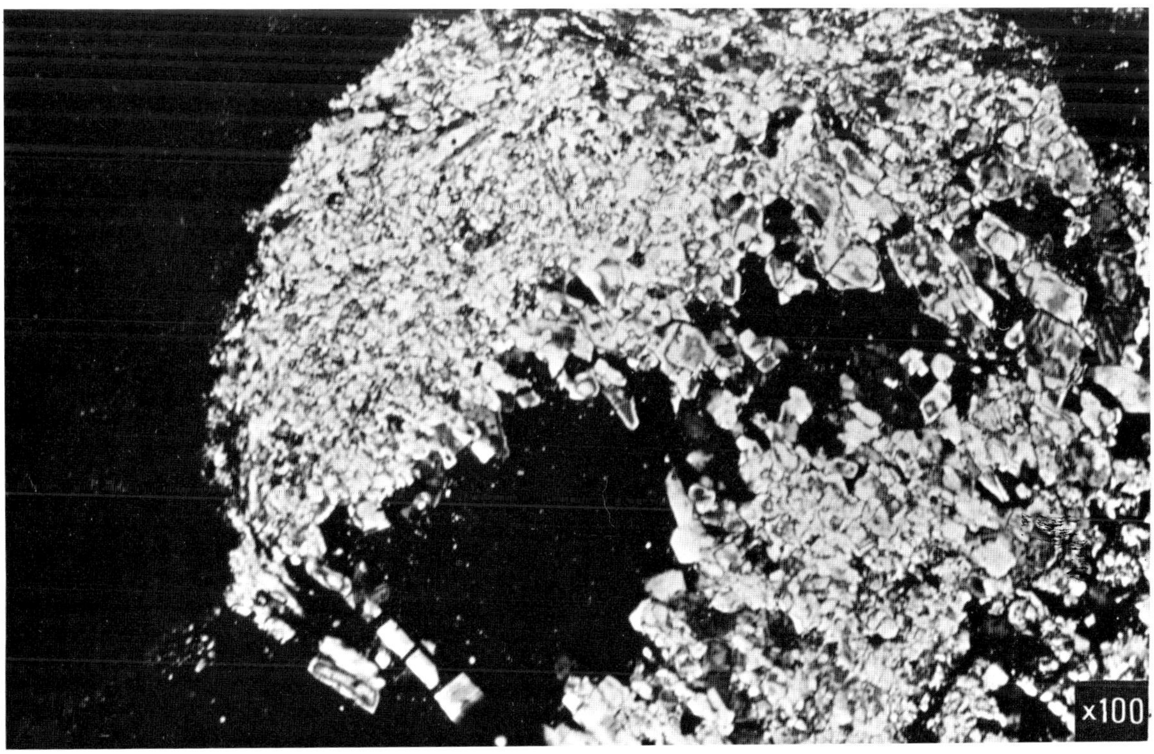

2

CHAPITRE VII (Planches 45 à 64). *CHAPTER VII (Plates 45 to 64)*

Eléments de pétrographie des évaporites oligocènes des bassins de la Bresse et de Valence (Est de la France, vallée de la Saône et du Rhône).

Petrographical data of Oligocene evaporites in the Bresse and Valence Basins.

PLANCHE 51 PSEUDOMORPHOSES DE GYPSE

Fig. 1 Pseudomorphose d'anhydrite à partir du gypse par transformation directe.

Fig. 2 et 3 Recristallisation de l'anhydrite après dissolution du gypse dont on devine encore les formes cristallines initiales.

PLATE 51 PSEUDOMORPHOSE OF GYPSUM

Fig. 1 Anhydrite pseudomorphose after gypsum, by direct transformation.

Figs. 2 and 3 Recrystallisation of anhydrite after dissolution of gypsum, in which original crystalline forms can still be divined.

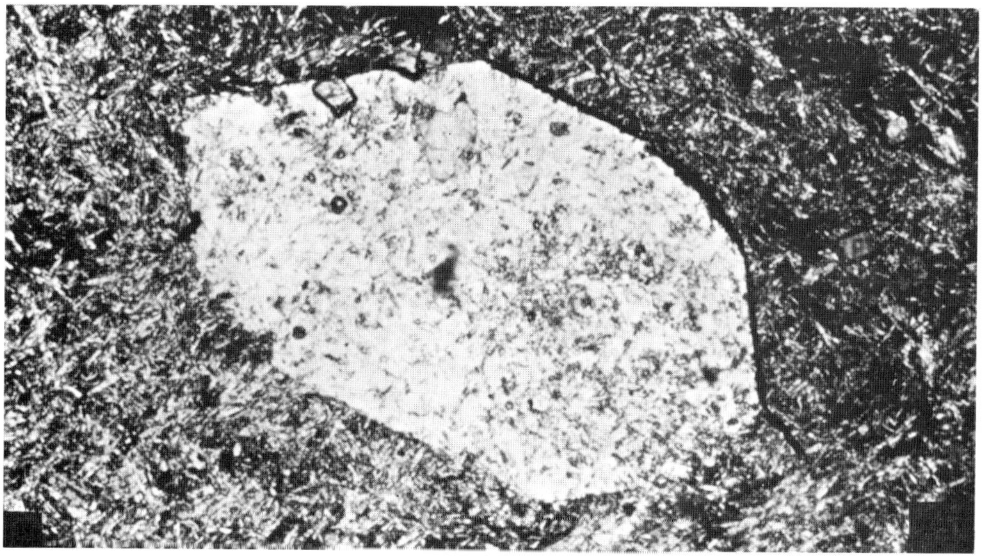

1

× 55,25

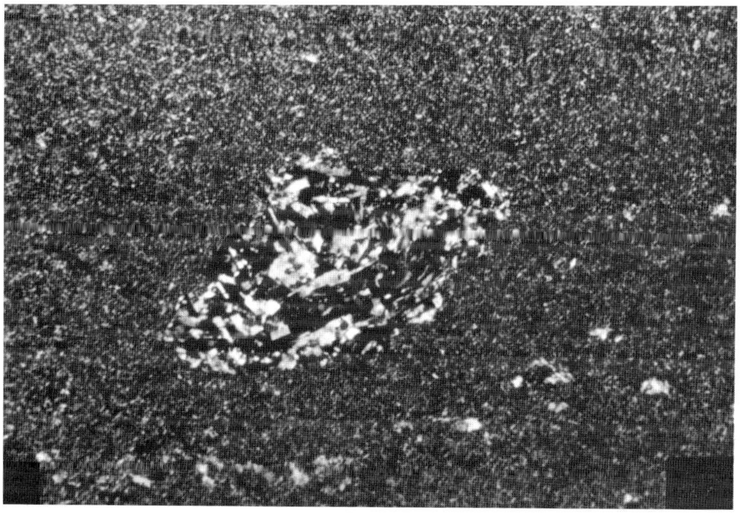

2

× 29,75

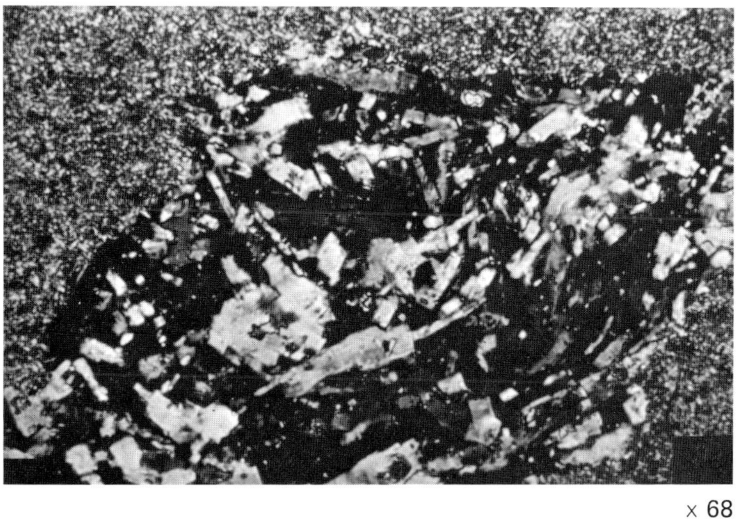

3

× 68

CHAPITRE VII (Planches 45 à 64). *CHAPTER VII (Plates 45 to 64)*

Eléments de pétrographie des évaporites oligocènes des bassins de la Bresse et de Valence (Est de la France, vallée de la Saône et du Rhône).

Petrographical data of Oligocene evaporites in the Bresse and Valence Basins.

PLANCHE 52 CROISSANCE DE NODULES D'ANHYDRITE

Fig. 1 Déformation de lits argilo-carbonatés au cours de la croissance de nodules d'anhydrite.

Fig. 2 Nodule à structure entérolithique.

PLATE 52 GROWING OF NODULAR ANHYDRITE

Fig. 1 Distortion of shaly-carbonated layers during growth on anhydrite nodules.

Fig. 2 Nodule with enterolithic structure.

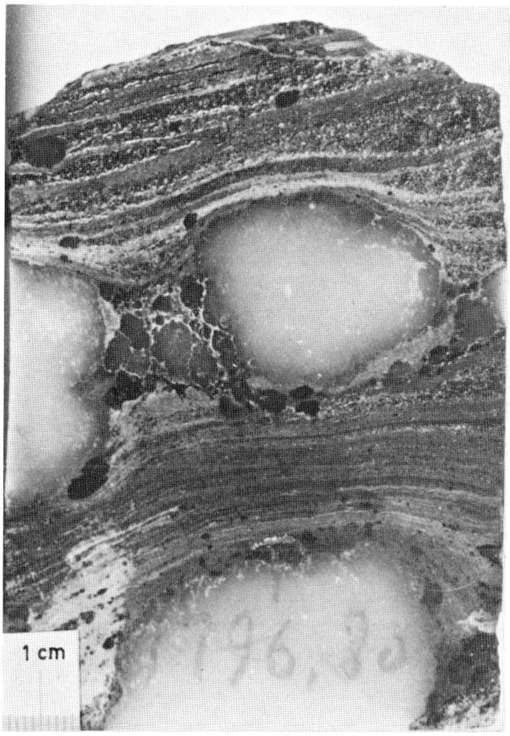

1

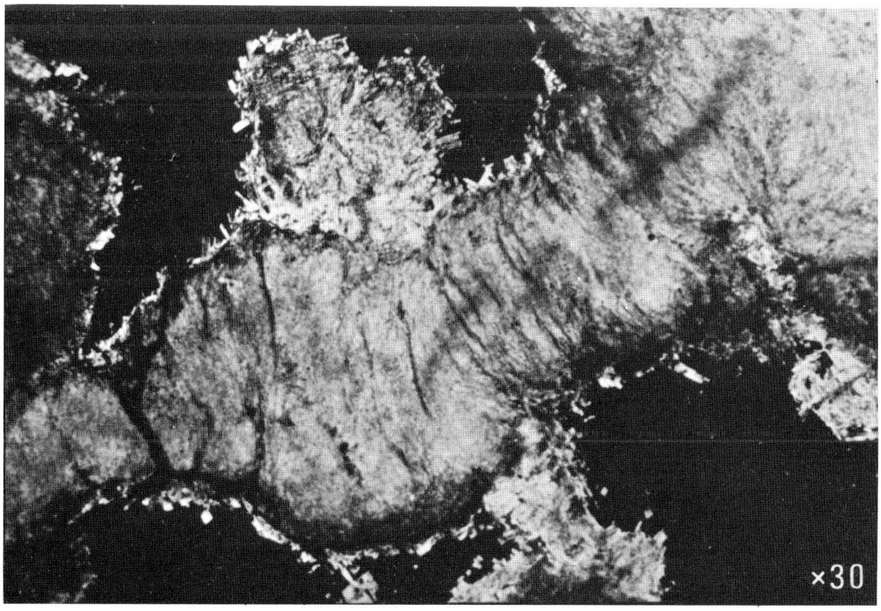

2

PLANCHE 53 ANHYDRITE EN MOSAIQUE

Fig. 1 Anhydrite à structure en mosaïque.

Fig. 2 Cristaux aciculaires perpendiculaires aux bords externes du nodule d'anhydrite.

PLATE 53 MOSAIC ANHYDRITE

Fig. 1 Anhydrite with mosaic structure.

Fig. 2 Perpendicular acicular crystals on external edges of the anhydrite nodule.

1

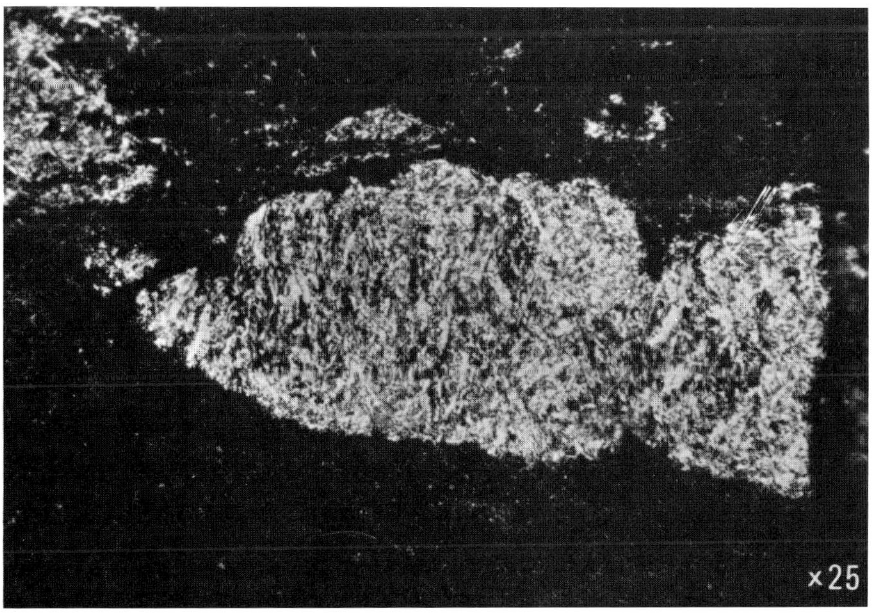

2

CHAPITRE VII (Planches 45 à 64). *CHAPTER VII (Plates 45 to 64)*

Eléments de pétrographie des évaporites oligocènes des bassins de la Bresse et de Valence (Est de la France, vallée de la Saône et du Rhône).

Petrographical data of Oligocene evaporites in the Bresse and Valence Basins.

PLANCHE 54 ANHYDRITE LAMINEE

PLATE 54 LAMINATED ANHYDRITE

1

2

PLANCHE 55 ANHYDRITE A STRUCTURE MAMELONNEE

PLATE 55 ANHYDRITE WITH MAMMILLATED STRUCTURE

1 cm

CHAPITRE VII (Planches 45 à 64). *CHAPTER VII (Plates 45 to 64)*

Eléments de pétrographie des évaporites oligocènes des bassins de la Bresse et de Valence (Est de la France, vallée de la Saône et du Rhône).

Petrographical data of Oligocene evaporites in the Bresse and Valence Basins.

PLANCHE 56 SILICE ASSOCIEE

Fig. 1 Mégaquartz en remplissage de cavité.

Fig. 2 Quartz au microscope électronique à balayage.

Fig. 3 Niveau détritique formé de mégaquartz, de plagioclases et de micas dans une matière argileuse riche en matière organique.

PLATE 56 ASSOCIATED SILICE

Fig. 1 Megaquartz filling a cavity.

Fig. 2 Quartz through scanning electron microscope.

Fig. 3 Detritic layer made up of megaquartz, plagioclases and micas in clay rich in organic matter.

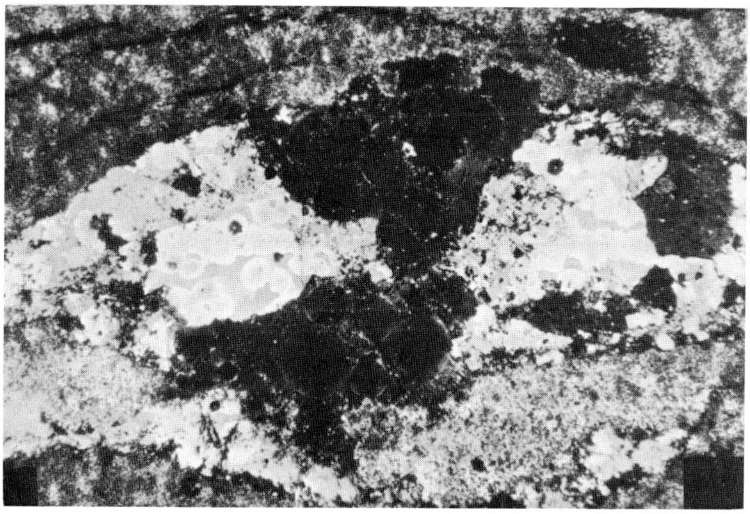

1

× 34

2

× 5100

3

× 29,75

CHAPITRE VII (Planches 45 à 64). *CHAPTER VII (Plates 45 to 64)*

Eléments de pétrographie des évaporites oligocènes des bassins de la Bresse et de Valence (Est de la France, vallée de la Saône et du Rhône).

Petrographical data of Oligocene evaporites in the Bresse and Valence Basins.

PLANCHE 57 SILICE ASSOCIEE

Fig. 1 et 2 Lit d'environ 1 cm de calcédoine massive à structure fibro-radiaire (fig. 1). Fantôme de structure nodulaire (fig. 2).

Fig. 3 Plage de microquartz à extinction ponctuelle dans de l'anhydrite massive.

PLATE 57 ASSOCIATED SILICE

Figs. 1 and 2 Massive chalcedony layer of approximately 1 cm with fibro-radiary structure (Fig. 1). Ghost of nodular structure (Fig. 2).

Fig. 3 Microquartz patch with punctual extinction in the massive anhydrite.

2

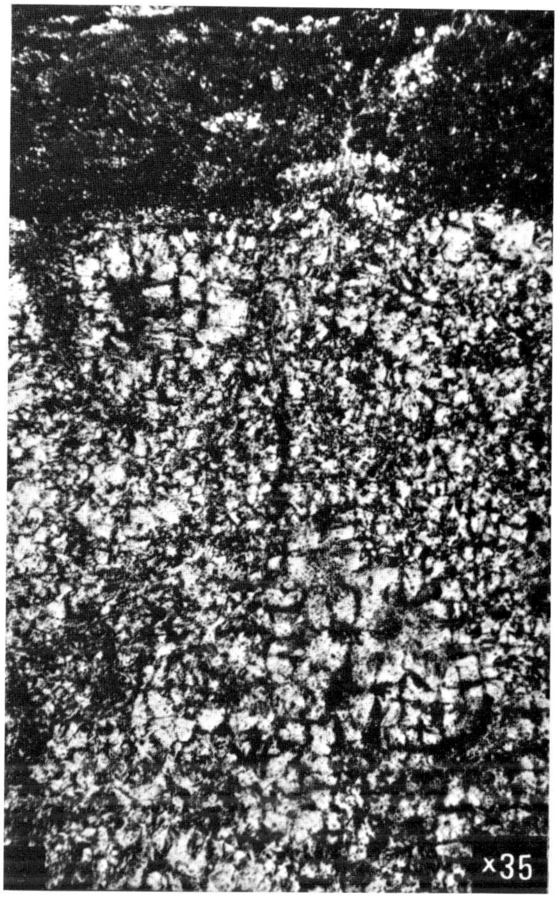

1

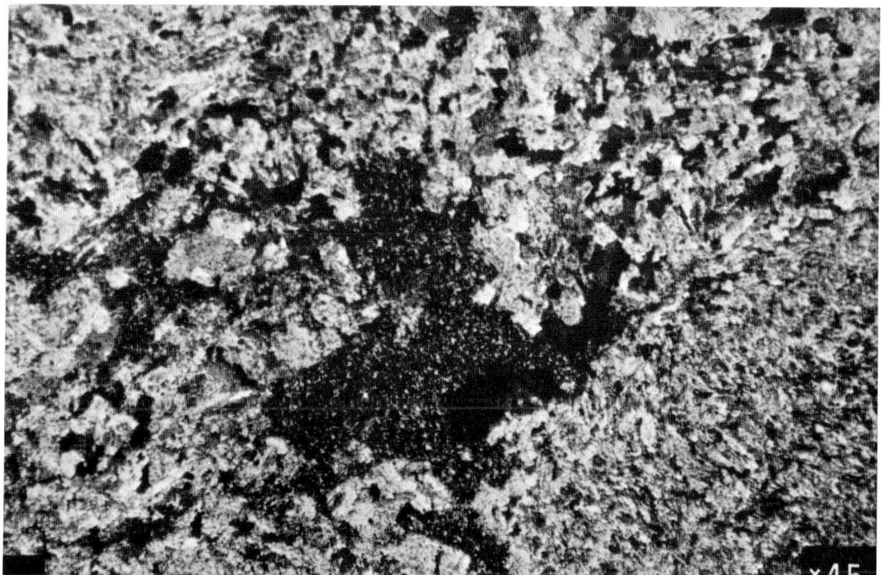

3

CHAPITRE VII (Planches 45 à 64). *CHAPTER VII (Plates 45 to 64)*

Eléments de pétrographie des évaporites oligocènes des bassins de la Bresse et de Valence (Est de la France, vallée de la Saône et du Rhône).

Petrographical data of Oligocene evaporites in the Bresse and Valence Basins.

PLANCHE 58 SILICE ASSOCIEE

Fig. 1 Remplacement de l'anhydrite par la silice dans une cavité.

Fig. 2 Mélange intime d'anhydrite (nodule) et de calcédoine.

PLATE 58 ASSOCIATED SILICE

Fig. 1 Replacement of anhydrite by silice in a cavity.

Fig. 2 Complete blend of anhydrite (nodule) and chalcedony.

1

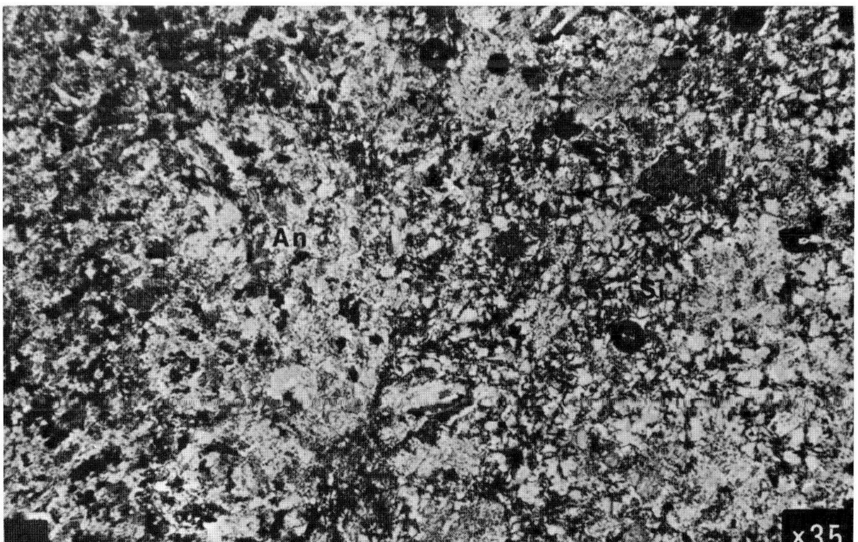

2

PLANCHE 59 LAMINITES ARGILO-CARBONATEES

Fig. 1 et 2 Laminites : dominance de teinte claire ou sombre selon la richesse en matière organique ou en carbonates.

PLATE 59 ARGILO-CARBONATE LAMINITES

Figs. 1 and 2 Laminites : predominantly lighter or darker in colour depending on the concentration of organic matter or carbonates respectively.

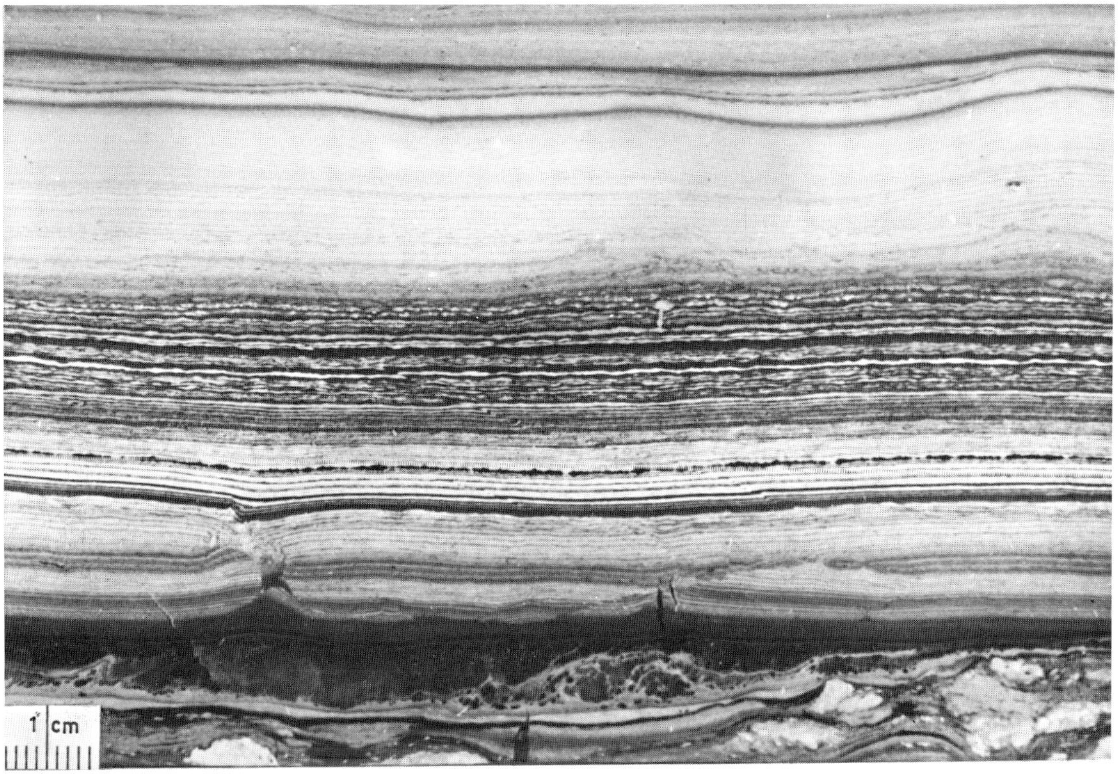

1

2

PLANCHE 60 COCCOLITHES DES LAMINITES

Fig. 1 *Reticulofenestra* sp. recouverts d'argile salifère.

Fig. 2 Calcaires dolomitiques à *Reticulofenestra* sp.

Fig. 3 *Braarudosphaera bigelowi* associé à *Reticulofenestra* sp.

Fig. 4 *Cruciplacolithes* sp.

PLATE 60 COCCOLITHS OF THE LAMINITES

Fig. 1 *Reticulofenestra* sp. overlapped by saliferous clay.

Fig. 2 Dolomitic limestones with *Reticulofenestra* sp.

Fig. 3 *Braarudosphaera bigelowi* coupled with *Reticulofenestra* sp.

Fig. 4 *Cruciplacolithes* sp.

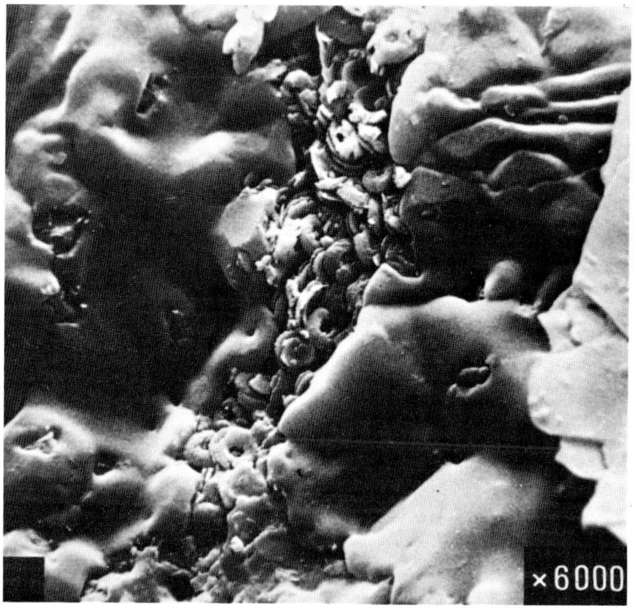

2

4

CHAPITRE VII (Planches 45 à 64). *CHAPTER VII (Plates 45 to 64)*

Eléments de pétrographie des évaporites oligocènes des bassins de la Bresse et de Valence (Est de la France, vallée de la Saône et du Rhône).

Petrographical data of Oligocene evaporites in the Bresse and Valence Basins.

PLANCHE 61 CYCLE EVAPORITIQUE MINEUR DU BASSIN DE BRESSE

Fig. 1 Petits nodules d'anhydrite répartis en lits horizontaux et en amas dans une matrice argilo-carbonatée.

Fig. 2 Nodule d'anhydrite de taille plus importante dans un lit d'argiles beiges.

Fig. 3 Anhydrite à structure entérolithique.

Fig. 4 Argiles plus ou moins carbonatées contenant des nodules d'anhydrite de taille différente.

Fig. 5 Laminites formées par une alternance de niveaux argilo-carbonatés (clairs) et argileux riches en matière organique (sombres).

PLATE 61 MINOR EVAPORITIC CYCLE OF THE BRESSE BASIN

Fig. 1 Small anhydrite nodules in horizontal layers and in concentration in a shaly-carbonated ground mass.

Fig. 2 Larger anhydrite nodule in beige-couloured clay layers.

Fig. 3 Anhydrite with enterolithic structure.

Fig. 4 More or less carbonate clays containing anhydrite nodules of a different size.

Fig. 5 Laminites generated by alternation of shaly-carbonated layers (light coloured) with darker clays layers rich in organic matter.

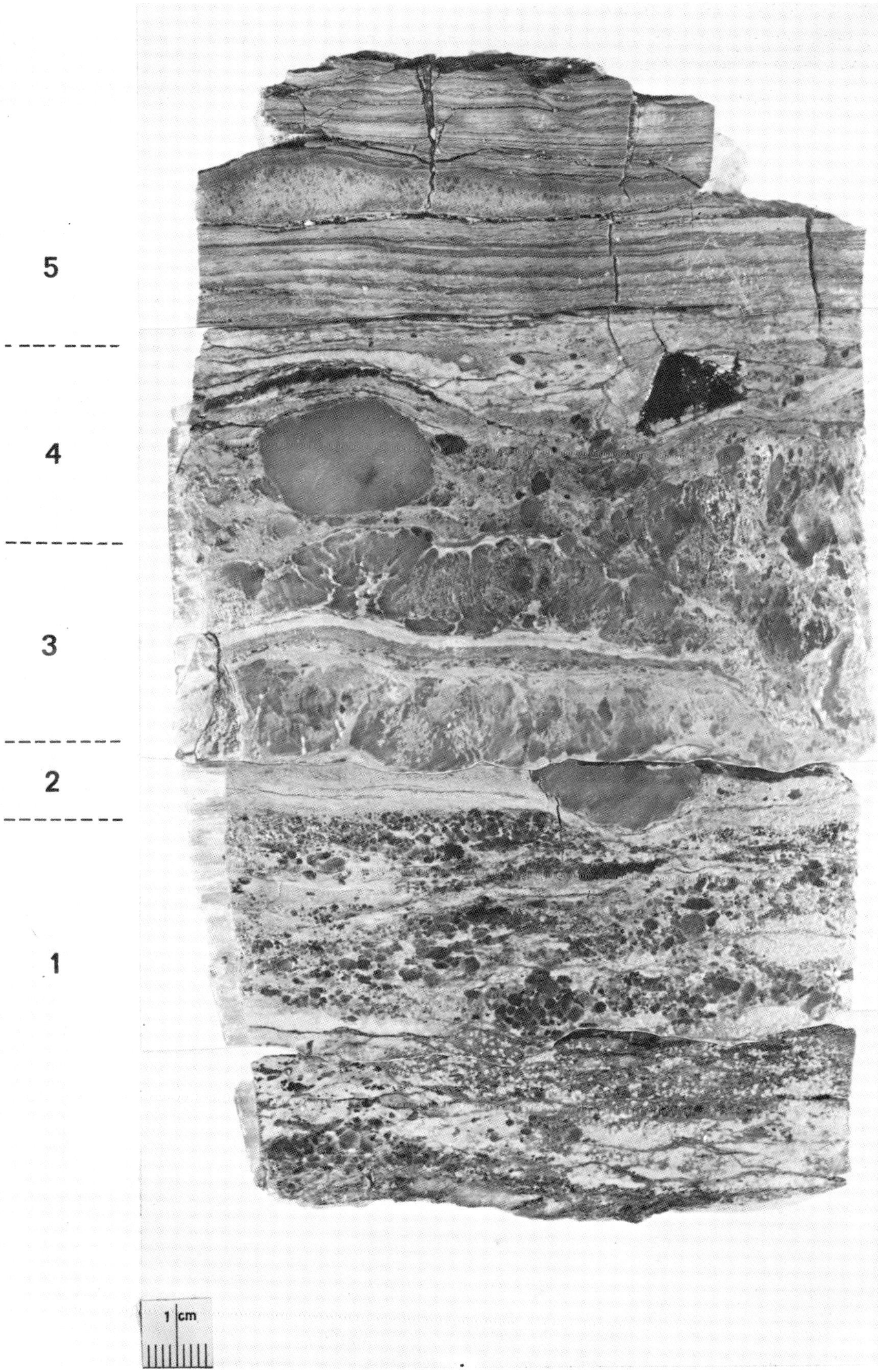
5
4
3
2
1
1 cm

CHAPITRE VII (Planches 45 à 64). *CHAPTER VII (Plates 45 to 64)*

Eléments de pétrographie des évaporites oligocènes des bassins de la Bresse et de Valence (Est de la France, vallée de la Saône et du Rhône).

Petrographical data of Oligocene evaporites in the Bresse and Valence Basins.

PLANCHE 62 CYCLE EVAPORITIQUE MINEUR DU BASSIN DE VALENCE

Fig. 1 et 2 Laminites argilo-carbonatées. Plus ou moins riches en matière organique.

Fig. 3 Niveau de gypses lenticulaires dissous, partiellement recristallisés en anhydrite.

Fig. 4 Argiles riches en matière organique renfermant des lits d'anhydrite nodulaire (planche 64, fig. 2).

Fig. 5 Anhydrite en mosaïque.

PLATE 62 MINOR EVAPORITIC CYCLE OF THE VALENCE BASIN

Figs. 1 and 2 Argilo-carbonated laminites. More or less rich in organic matter.

Fig. 3 Layer of dissolved lenticular gypsum, partially recrystallised into anhydrite.

Fig. 4 Clays rich in organic matter including layers of nodular anhydrite (Plate 64, Fig. 2).

Fig. 5 Mosaic anhydrite.

5

4

3

2

1

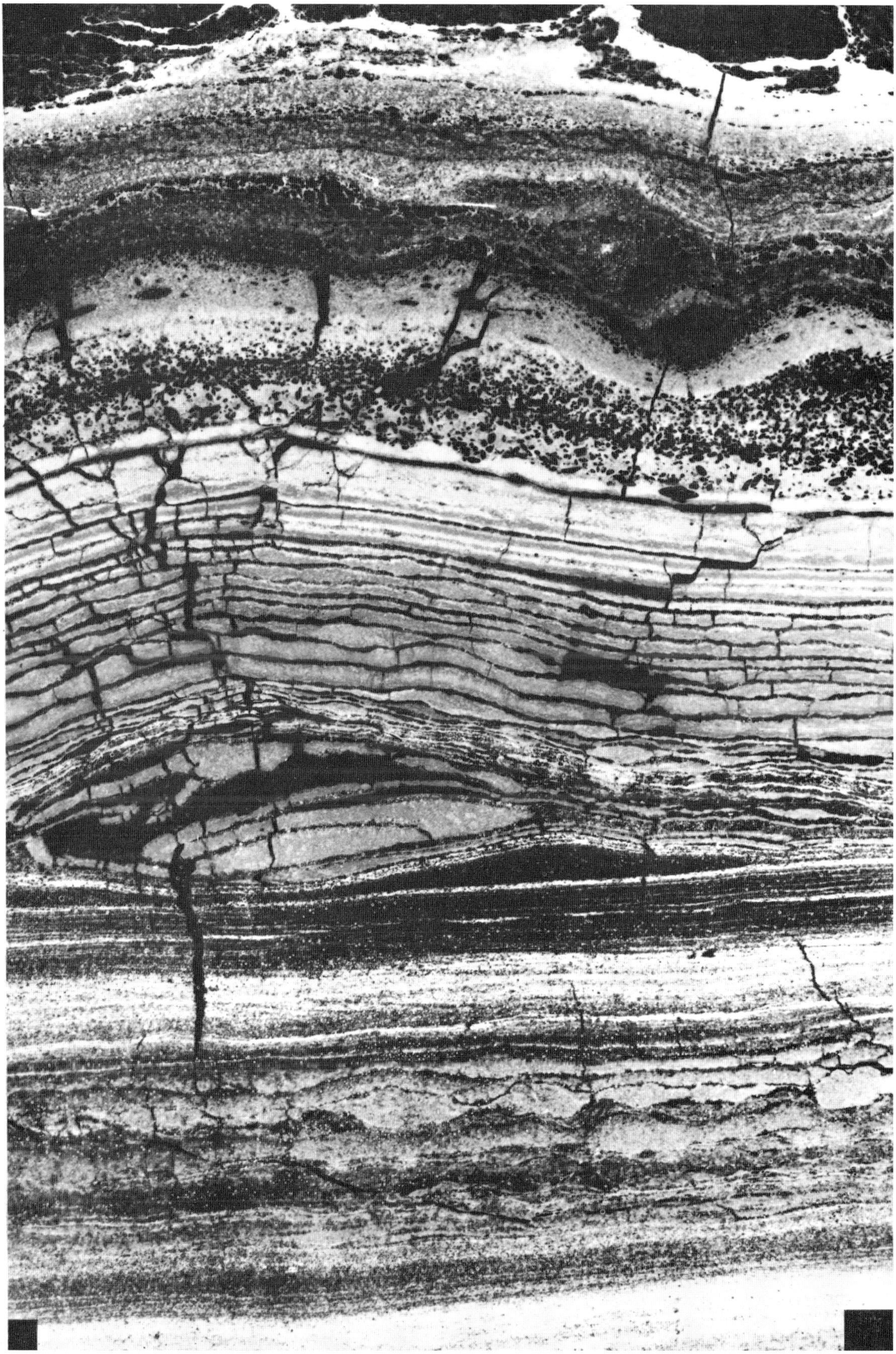

× 6,8

CHAPITRE VII (Planches 45 à 64). *CHAPTER VII (Plates 45 to 64)*

Eléments de pétrographie des évaporites oligocènes des bassins de la Bresse et de Valence (Est de la France, vallée de la Saône et du Rhône).

Petrographical data of Oligocene evaporites in the Bresse and Valence Basins.

PLANCHE 63 DISSOLUTION-RECRISTALLISATION DE GYPSE

Fig. 1 et 2 Dissolution de gypse lenticulaire, puis recristallisation partielle ou totale en anhydrite microcristalline.

PLATE 63 SOLUTION AND RECRYSTALLISATION OF GYPSUM

Figs. 1 and 2 Dissolution of lenticular gypsum, then partial or total recrystallisation into microcrystalline anhydrite.

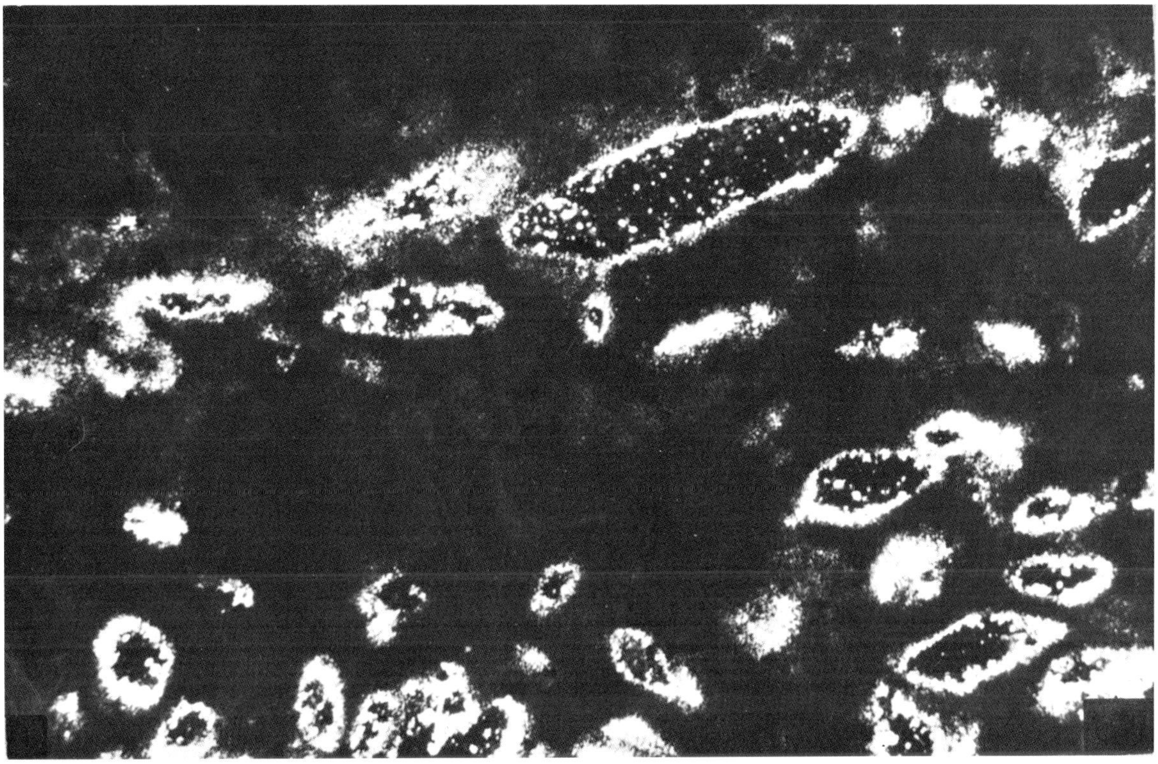

1

x 38

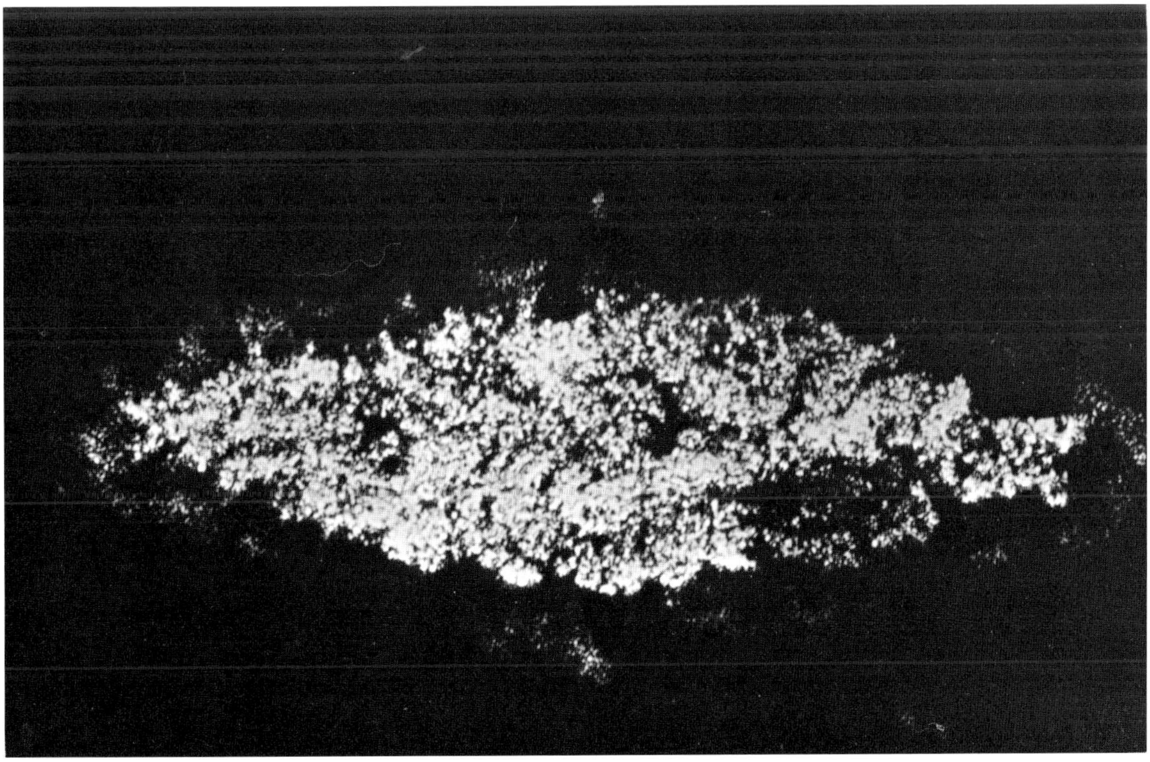

2

x 90,25

CHAPITRE VII (Planches 45 à 64). *CHAPTER VII (Plates 45 to 64)*

Eléments de pétrographie des évaporites oligocènes des bassins de la Bresse et de Valence (Est de la France, vallée de la Saône et du Rhône).

Petrographical data of Oligocene evaporites in the Bresse and Valence Basins.

PLANCHE 64 DETAILS DE LA SEQUENCE DU BASSIN DE VALENCE

Fig. 1 Cet échantillon représente le cycle sédimentaire de la planche 62 reposant sur un banc d'anhydrite massive, et surmonté par d'autres dépôts carbonatés.

Fig. 2 Lits d'anhydrite nodulaire.

PLATE 64 DETAILS OF THE SEQUENCE FROM THE VALENCE BASIN

Fig. 1 This sample represents the sedimentary cycle of Plate 62 overlying a bed of massive anhydrite and overbedded by other carbonated deposits.

Fig. 2 Beds of nodular anhydrite.

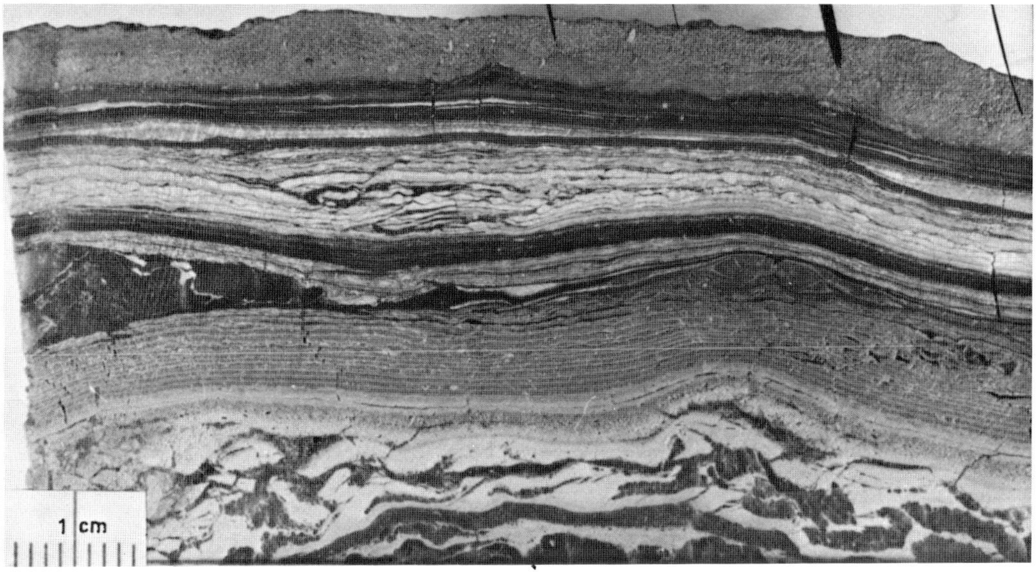

1

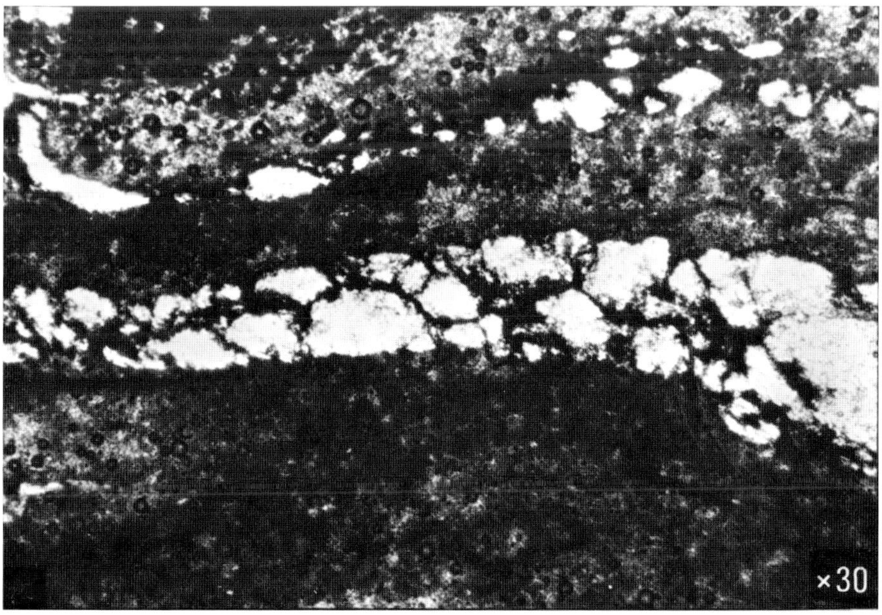

2

CHAPITRE VIII (Planches 65 à 69). *CHAPTER VIII (Plates 65 to 69)*

Anhydrite diagénétique de sebkha carbonatée dans le Tertiaire d'Irak.

Diagenetic anhydrite of carbonated sebkha in the Tertiary of Iraq.

PLANCHE 65 DEPOTS ALGAIRES DE SEQUENCE DE LAGON

Fig. 1 *Boundstone* algaire dolomitisé.

Fig. 2 *Boundstone* algaire dolomitisé et remanié.

PLATE 65 ALGAL DEPOSITS IN A LAGOONAL SEQUENCE

Fig. 1 Dolomitized algal boundstone.

Fig. 2 Reworked dolomitized algal boundstone.

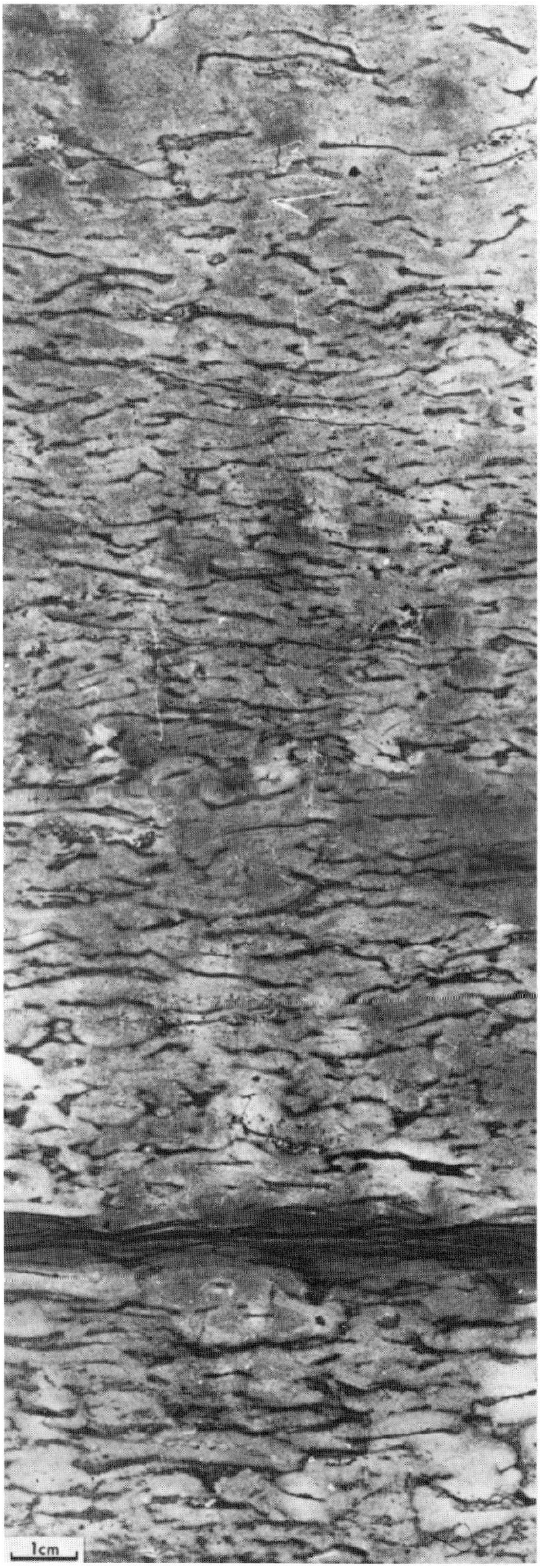

1

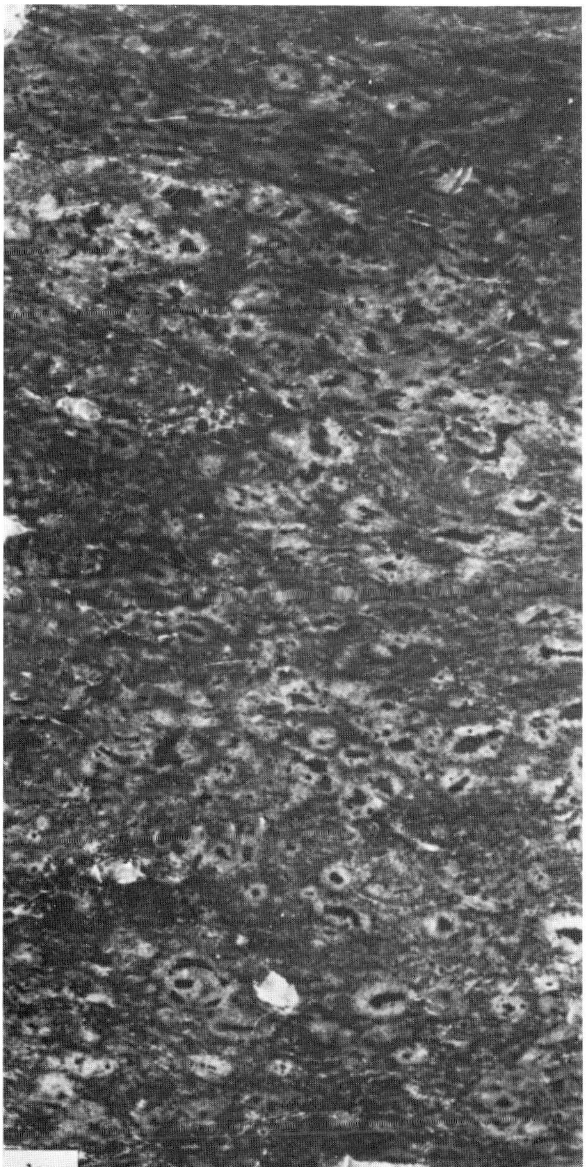

2

CHAPITRE VIII (Planches 65 à 69). *CHAPTER VIII (Plates 65 to 69)*

Anhydrite diagénétique de sebkha carbonatée dans le Tertiaire d'Irak.

Diagenetic anhydrite of carbonated sebkha in the Tertiary of Iraq.

PLANCHE 66 FACIES MACROSCOPIQUES D'ANHYDRITE DIAGENETIQUE

Fig. 1 Anhydrite en microplages : L'anhydrite est répartie en fines plages avec fréquemment présence de cristaux pseudomorphes soit dans une dolomicrite, soit dans un dépôt *wackestone* dolomitisé à Foraminifères et Ostracodes.

Fig. 2 Faciès pseudobréchique : Quelques nodules d'anhydrite *chicken meshs* apparaissent dans un faciès pseudobréchique comportant d'abondants nodules dolomitiques de même morphologie que les *chicken meshs* avec remplacement probable de l'anhydrite par de la dolomie. La succession des phénomènes a dû être la suivante :

a) Dépôt d'un *boundstone* algaire.
b) Dolomitisation des algues.
c) Formation d'anhydrite nodulaire.
d) Remplacement partiel de l'anhydrite.
e) Bréchification.

PLATE 66 MACROSCOPIC FACIES OF DIAGENETIC ANHYDRITE

Fig. 1 Micropatches of anhydrite : These micropatches show pseudomorphe crystals intradolomicrite or into a dolomitic wackestone with Foraminifera and Ostracodes.

Fig. 2 Pseudobrecciated facies : Some chicken meshs anhydritic nodules appear into a pseudobrecciated facies comprising abundant dolomitic nodules of the same morphology as chicken meshs with a probable replacement of anhydrite by dolomite.
The succession of phenomenons has been the following one :

a) Deposit of an algal boundstone.
b) Dolomitisation of the algal material.
c) Nodular anhydrite formation.
d) Partial replacement of anhydrite.
e) Brecciaes formation.

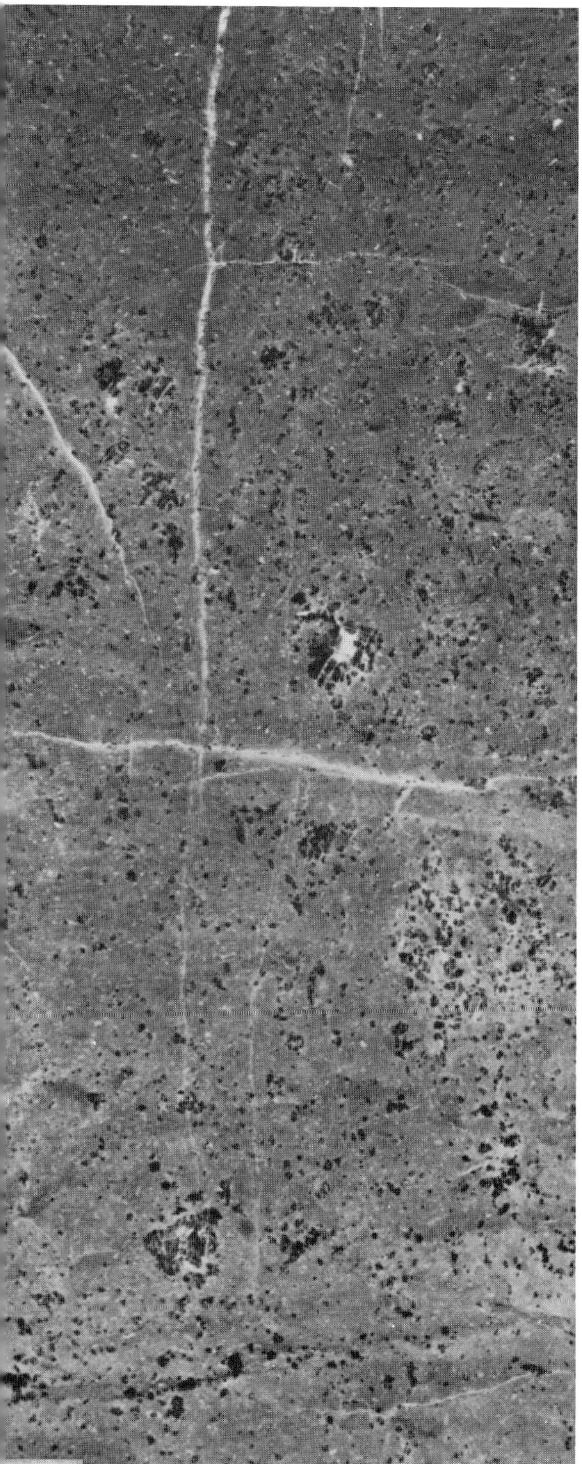
1cm

1

Bréchification
Brecciated
1cm
Boundstone algaire
Algal boundstone

2

CHAPITRE VIII (Planches 65 à 69). *CHAPTER VIII (Plates 65 to 69)*

Anhydrite diagénétique de sebkha carbonatée dans le Tertiaire d'Irak.

Diagenetic anhydrite of carbonated sebkha in the Tertiary of Iraq.

PLANCHE 67 FACIES MACROSCOPIQUES D'ANHYDRITE DIAGENETIQUE

Fig. 1 Anhydrite nodulaire (*chicken meshs*) : Dans un faciès algaire les nodules d'anhydrite se développent ; remarquer leur concentration croissante du haut vers le bas et leur disparition brutale en-dessous d'une surface pratiquement horizontale.

Fig. 2 Anhydrite irrégulièrement massive : L'anhydrite ne se présente plus en nodules mais semble envahir une grande partie de faciès dolomitisés à microforaminifères et pellets. Des fantômes de dolomie subsistent. Comme dans le cas de *chicken meshs,* la concentration d'anhydrite croît vers la base et disparaît brutalement en-dessous d'une surface pratiquement horizontale.

PLATE 67 MACROSCOPIC FACIES OF DIAGENETIC ANHYDRITE

Fig. 1 Nodular anhydrite (chicken meshs) : Nodules of anhydrite are growing into an algal facies. See their increasing towards the bottom and their striking disappearance beneath an almost horizontal surface.

Fig. 2 Irregular massive anhydrite : It is not a nodular anhydrite but it seems mostly developed into the dolomitised facies with pellets and microforaminifera. Dolomitic ghosts remain. As in chicken meshs case, the anhydrite concentration increases towards the bottom, and strikingly disappears beneath an almost horizontal surface.

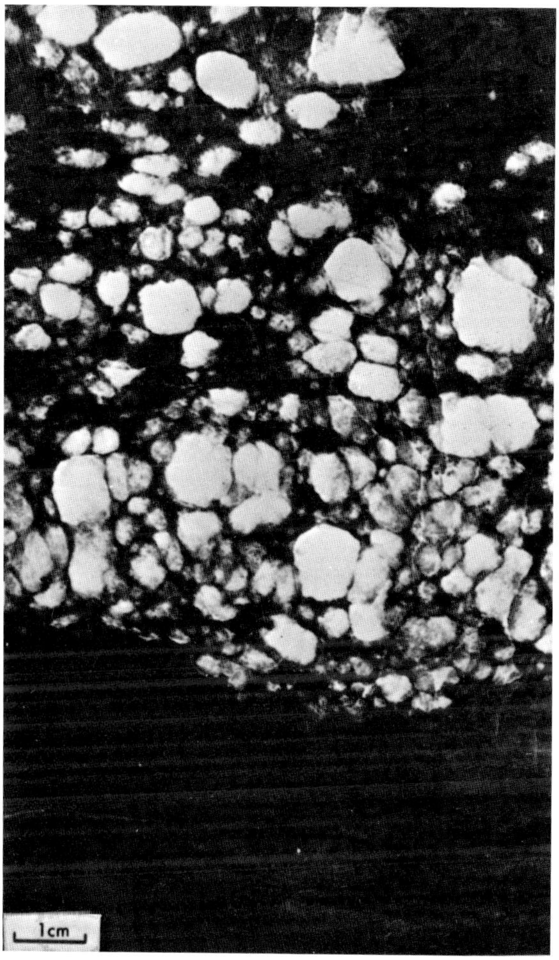

1

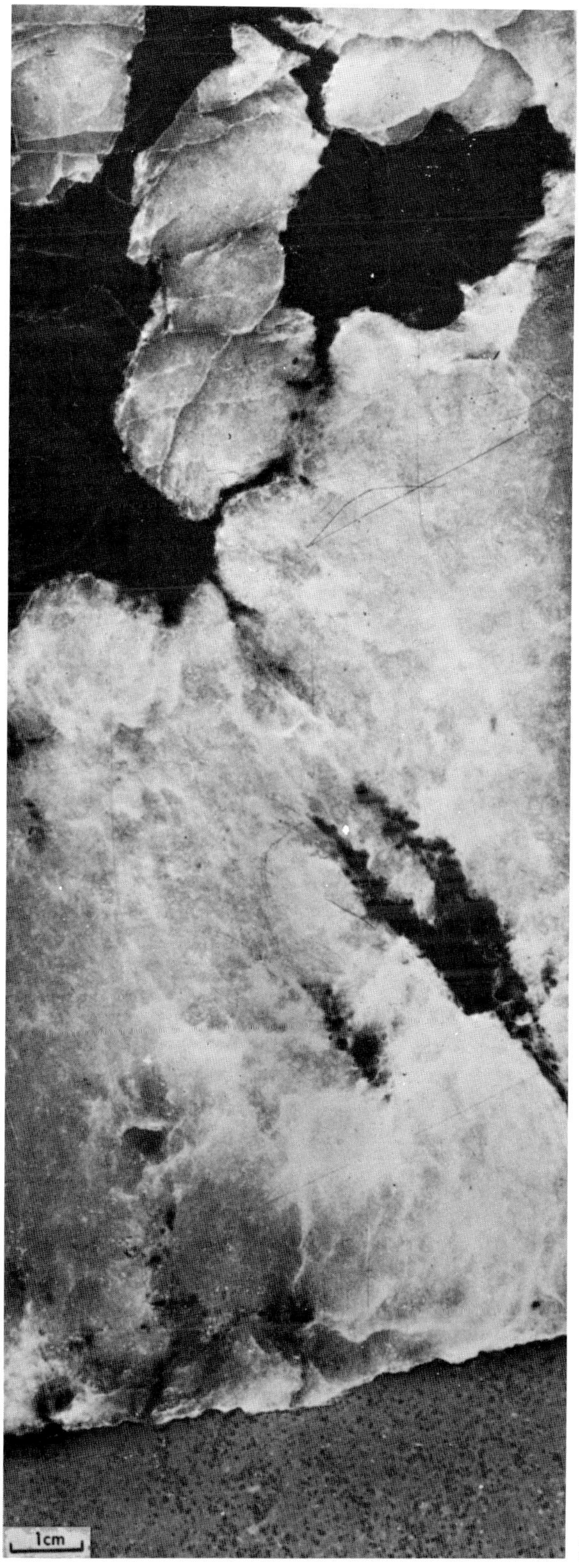

2

CHAPITRE VIII (Planches 65 à 69). *CHAPTER VIII (Plates 65 to 69)*

Anhydrite diagénétique de sebkha carbonatée dans le Tertiaire d'Irak.

Diagenetic anhydrite of carbonated sebkha in the Tertiary of Iraq.

PLANCHE 68 MICROFACIES D'ANHYDRITE DIAGENETIQUE

Fig. 1 Anhydrite de remplacement : Des monocristaux d'anhydrite remplacent la dolomie, on peut apercevoir les débris fantômatiques de dolomie dans l'anhydrite.

Fig. 2 Anhydrite de remplacement : On peut apercevoir le détail d'un remplacement. Un fantôme de Foraminifères *(Miliolidae)* apparaît très nettement ; il n'y a donc pas eu dissolution d'une plage et remplissage mais remplacement uniquement.

Fig. 3 Anhydrite de remplacement : Contrairement à l'exemple précédent, la plage du monocristal d'anhydrite apparaît très claire, pratiquement sans inclusions, il y a donc eu dissolution vacuolaire (on aperçoit une vacuole non remplie à droite de la figure), remplissage anhydritique et remplacement sur les bords de la vacuole (voir auréole à inclusions).

PLATE 68 DIAGENETIC ANHYDRITE MICROFACIES

Fig. 1 Replacement anhydrite : Anhydrite monocrystals replaced the dolomite. We can see dolomitic ghosts in the anhydrite.

Fig. 2 Replacement anhydrite : We can see a detail of the replacement. A Foraminiferal ghost clearly appears. This is not a patch dissolution and infilling but only a replacement.

Fig. 3 Infilling anhydrite : On the contrary with Fig. 2, the monocrystal anhydrite patch appears very clear, without inclusions, this is a vuggy dissolution (we can see an empty vug on the right of the figure), then anhydritic infilling and replacement on the borders of the vug (see inclusions on the borders).

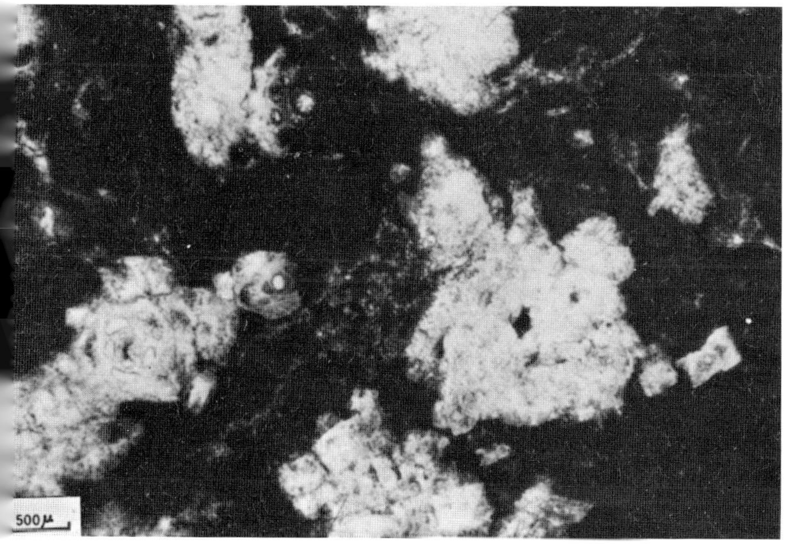

1

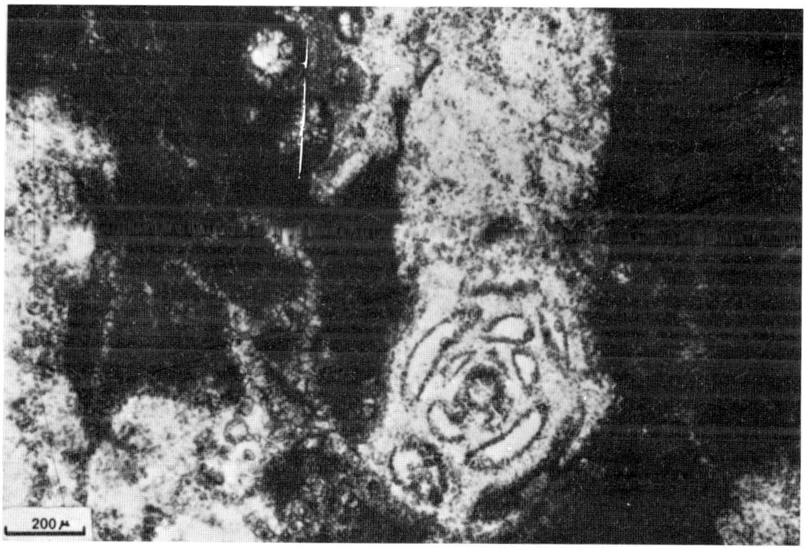

2

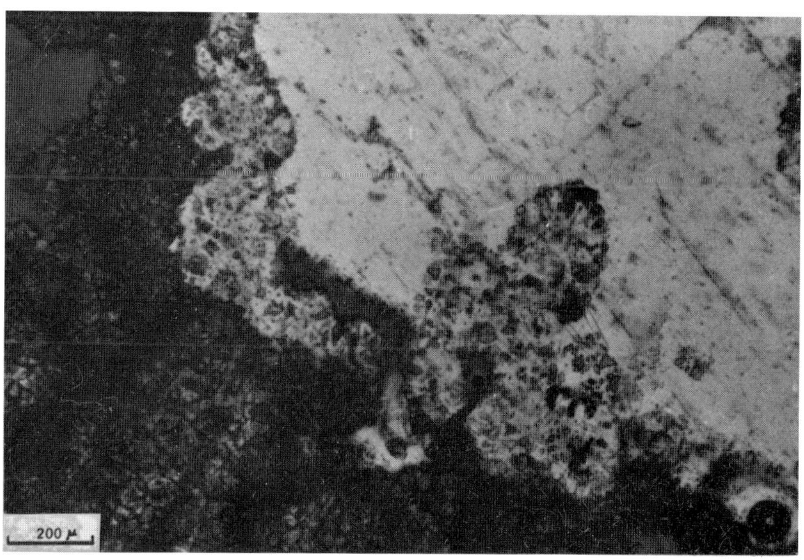

3

CHAPITRE VIII (Planches 65 à 69). *CHAPTER VIII (Plates 65 to 69)*

Anhydrite diagénétique de sebkha carbonatée dans le Tertiaire d'Irak.

Diagenetic anhydrite of carbonated sebkha in the Tertiary of Iraq.

PLANCHE 69 MICROFACIES D'ANHYDRITE DIAGENETIQUE

Fig. 1 Anhydrite nodulaire : On peut remarquer le contact franc entre l'anhydrite et la dolomie et la cristallisation en nombreuses aiguilles allongées dans des sens divers mais à partir de points centraux (voir schéma ci-dessous).

Fig. 2 Anhydrite de remplissage : L'anhydrite remplit ici une coquille dont ne subsiste que les parois externes, le remplissage se fait par un réseau de cristaux allongés analogues à ceux des nodules, donc d'origine probablement identique.

PLATE 69 DIAGENETIC ANHYDRITE MICROFACIES

Fig. 1 Nodular anhydrite : We can note the right contact between anhydrite and dolomite and the crystallisation into numerous needles. These needles are disposed in many directions since central points (see diagram).

Fig. 2 Infilling anhydrite : Here the anhydrite is filling a bioclasts whose external test is only present. The infilling is made by a network of crystals similar to those of the nodules, and probably of the same origin.

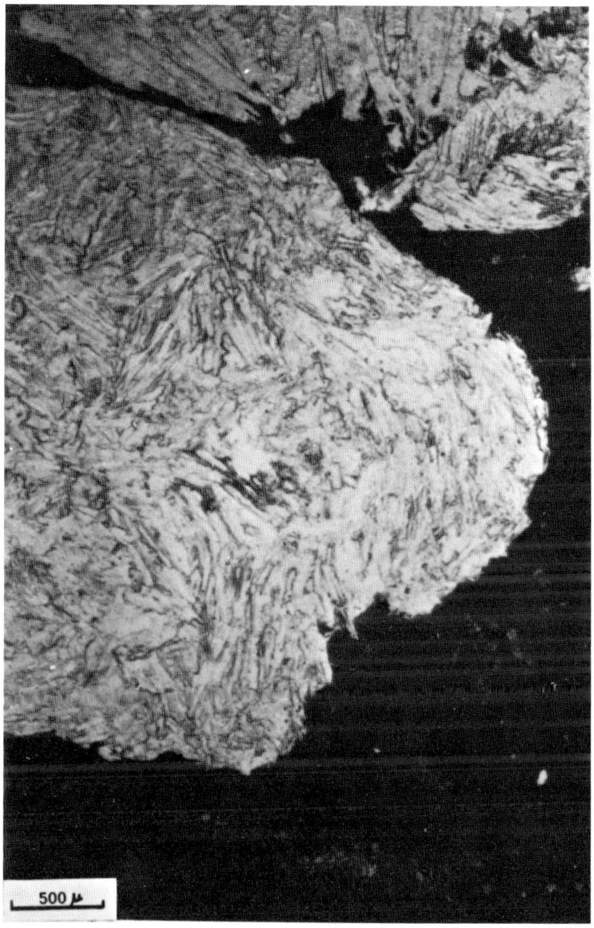

1

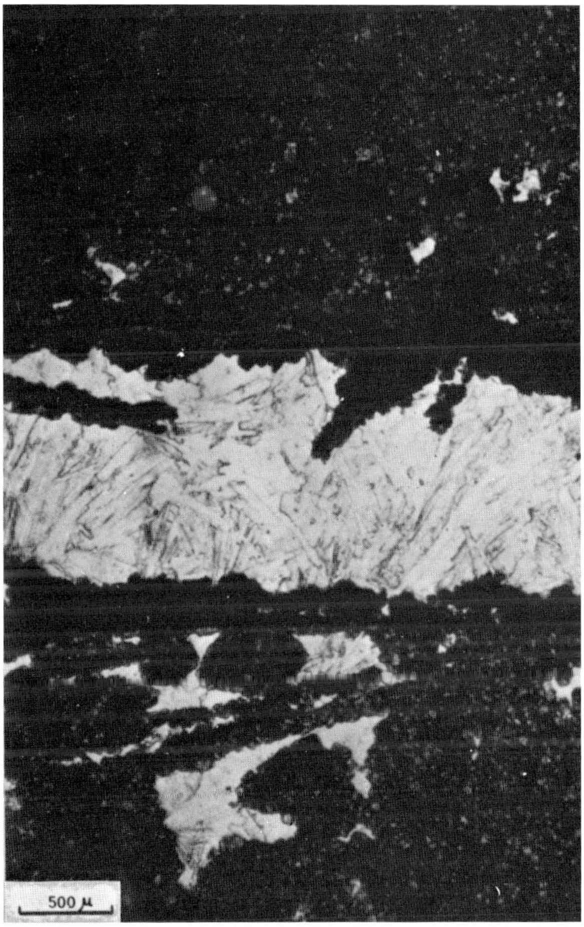

2

CHAPITRE IX (Planches 70 à 72). *CHAPTER IX (Plates 70 to 72)*

Habitats et diagenèse de gypses miocènes. Benejuzar (Alicante), Espagne.

Habitats and Miocene gypsum diagenesis. Benejuzar (Alicante), Spain.

PLANCHE 70 DOMES GYPSIFERES

Vues de quelques "dômes" gypsifères avec "gouttières" de sédiments laminés ; l'ensemble de la couche fait environ 10 m d'épaisseur. Constitués avant tout de colonnes de gypse sélénite en fer de lance, une première cause aux "gouttières" semble une morphologie vallonnée du sommet du gypse. Des lits de laminites semblent traverser horizontalement les espaces entre colonnes gypseuses, supposées primaires (voir Cabo-Sharman).

Cependant des nodules de gypse alabastrite, diagénétiques, puis des zones de gypse mosaïque (terme extrême), attestent une augmentation de volume qui fait rebrousser vers le haut les couches de carbonates laminés. Un pseudo-diapirisme (de cristallisation) est ainsi suggéré, et confirmé par l'attitude structurale des couches sus-jacentes (cristallisation et ré-hydratation d'anhydrite secondaire en gypse).

PLATE 70 GYPSIFEROUS "DOMES"

Views of some gypsiferous "domes" with laminated sediment "gutters" ; the whole layer is about 10 m thick. These "gutters" seem due to vale morphology of the top of gypsum made of mostly selenite swallow-tailed gypsum columns. Horizontal laminite layers cut through the spaces between the gypsum columns, supposed of primary origin (see Cabo-Shearman).

However diagenetic nodules of alabastrite gypsum and (at the extreme) zones of mosaic gypsum show an increase in volume causing an upturning of the laminated carbonate layers. A pseudo-diapirism (of crystallisation) is therefore suggested, and confirmed by the structure of underlying layers (crystallisation and rehydratation of secondary anhydrite into gypsum).

1

2

3

CHAPITRE IX (Planches 70 à 72). *CHAPTER IX (Plates 70 to 72)*

Habitats et diagenèse de gypses miocènes. Benejuzar (Alicante), Espagne.

Habitats and Miocene gypsum diagenesis. Benejuzar (Alicante), Spain.

PLANCHE 71 DETAILS D'UN MONTICULE GYPSEUX

Détails de l'un des monticules de gypse (chaque barre sombre représente un centimetre).

Vue des colonnes de gypse fer de lance, verticales, la pointe tournée vers le bas (règle "de Motura", voir Cabo Shearman).

Le détail du bas confirme l'invasion du gypse sélénite par le gypse nodulaire blanc (alabastrite) que l'on peut deviner sur la photo du haut, envahissant la colonne du haut-gauche. Sous la réglette, gypse nodulaire formant une mosaïque diagénétique.

La matrice semble composée surtout de matière organique sombre.

PLATE 71 GYPSUM MONTICULE DETAIL

A gypsum monticule in detail (each dark bar represents one centimetre).

View of vertical swallow-tailed gypsum columns, pointing downwards ("Motura law", see Cabo Shearman).

The lower structure confirms the invasion of selenite gypsum by white nodular gypsum (alabastrite), which can be identified in the top photograph, invading the top left column. Nodular gypsum is seen – under the ruler – to form a diagenetic mosaic.

The ground mass seems to be made up of dark coloured organic matter.

1

2

CHAPITRE IX (Planches 70 à 72). *CHAPTER IX (Plates 70 to 72)*

Habitats et diagenèse de gypses miocènes. Benejuzar (Alicante), Espagne.

Habitats and Miocene gypsum diagenesis. Benejuzar (Alicante), Spain.

PLANCHE 72 "RIPPLE MARKS" DANS LES COUCHES GYPSEUSES (découverts par C. Montenat).

L'origine sub-aquatique de ces couches paraît indéniable.

Cependant, la disposition des cristaux de gypse pose certains problèmes : les cristaux semblent déposés dans une matrice boueuse, et certains ont continué à pousser diagénétiquement ainsi que l'attestent les figures cristallines ennoyées dans l'argile sus-jacente (au creux de certains *ripples*).

L'allure dissymétrique de la granulométrie des cristaux, visible surtout sur la photographie du bas, semble être due à l'eau de ruissellement actuel qui a dissous, en coulant de gauche à droite, la face la plus exposée.

PLATE 72 RIPPLE MARKS IN GYPSIFEROUS LAYERS (discovered by C. Montenat).

The subaquatic origin of these layers seems definite.

However, the arrangement of the gypsum crystals causes certain problems : they seem to be deposited in a muddy ground mass, and some continued to expand diagenetically as borne out by crystalline shapes sunk into the underlying clay (in hollows of certain ripples).

The asymmetry of crystal granulometry, particularly evident in the bottom photograph, could be caused by present day rain wash, which, moving left to right, dissolved the most exposed side.

1

2

CHAPITRE X (Planches 73 et 74). *CHAPTER X (Plates 73 and 74)*

Habitats évaporitiques du Silurien du Michigan (Etats-Unis).

Evaporitic habitats of the Michigan Silurian (USA).

PLANCHE 73 TRANSITION ENTRE LES CARBONATES A_0 ET LES EVAPORITES A_1.

Fig. 1 Laminites mécaniques. Les couches claires correspondent aux carbonates anhydritiques, les couches grises, recessives (dissolution) correspondent à de la halite claire (de remplacement ?). Quelques cristaux de halite dans les couches claires. Vers la base le bombement clair et les biseautages latéraux (flèche) sont interprétés comme une figure d'expulsion d'eau intersticielle.
(Forage Paul et Trapp).

Fig. 2 Contact entre des carbonates d'origine algaire possible et des varves à anhydrite massive.
(Forage Paul et Trapp).

Fig. 3 Anhydrite en mosaïque avec des reliques de laminites soit digérées pendant la diagenèse soit originellement déposées (varves) dans une gouttière entre les cristaux de gypse ultérieurement remplacée par l'anhydrite. Cette ambiguïté illustre bien les difficultés dans les reconstitutions sédimentologiques, en milieu "hyperdiagénétique".
(Forage Shell Longfield).

Fig. 4 Un cycle ascendant carbonates-évaporites dans évaporite A_1.
Les deux morceaux de carottes de droite sont formés d'anhydrite massive à tendance "mosaïque". La base de la série carbonatée montre des figures contournées (avec interlits anhydritiques) attribuées à des slumps en milieu varvaire. Des raisons géométriques aident à éliminer l'hypothèse de figures contournées de couches algaires bien que l'incertitude persiste, en comparant avec les multiples formes que prennent les faciès algaires.
(Forage Shell Longfield).
Graduations en centimètres.

PLATE 73 TRANSITION BETWEEN A_0 CARBONATES AND A_1 EVAPORITES

Fig. 1 Mechanical laminites. The light coloured layers are of anhydritic carbonates, the recessive grey layers (dissolution) are of light halite (of replacement ?). There are some halite crystals in the light layers. Towards the base, light up-warping and lateral bevelling (boom) suggest expulsion of interstitial water.
(Paul and Trapp well).

Fig. 2 Contact between carbonates of possible algal origin and varves with massive anhydrite.
(Paul and Trapp well).

Fig. 3 Mosaic anhydrite with laminite remains either digested during diagenesis or originally deposited (varves) in the "gutter" between the gypsum crystals later replaced by anhydrite. This ambiguity exemplifies the difficulties encountered when reconstructing deposition in "hyperdiagenetic" conditions.
(Shell Longfield well).

Fig. 4 The cycle runs left to right from carbonates to evaporites in A_1 evaporites.
Two pieces of core from the right are of massive anhydrite, somewhat mosaic anhydrite. The base of the carbonated series shows contorted shapes (with anhydritic interlayers) attributed to slumps in varvary conditions – geometric explanations help to reject the hypothesis that these contortions are algal layers although this is not decisive, by comparing them with the various forms taken by algal facies.
(Shell Long field well).
Graduations in centimetres.

1

2

3

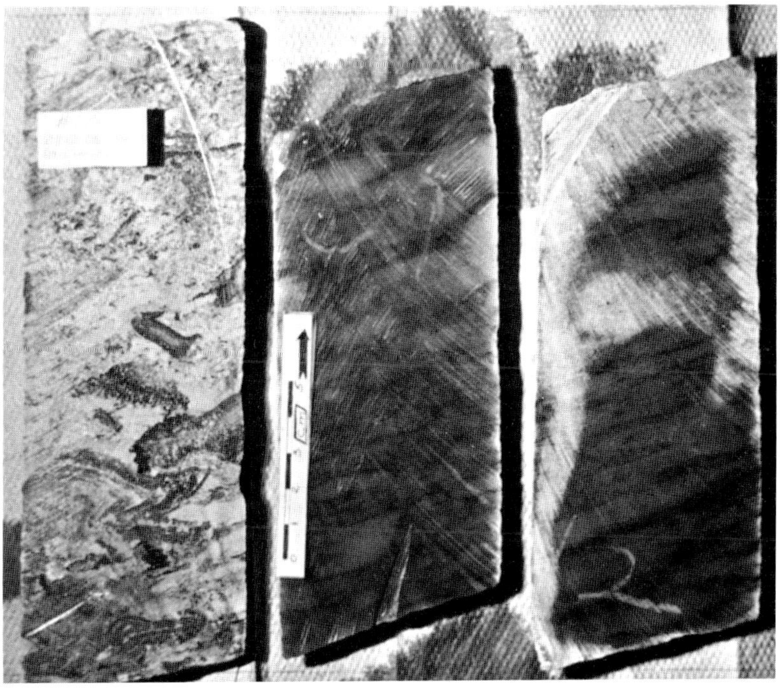
4

CHAPITRE X (Planches 73 et 74). *CHAPTER X (Plates 73 and 74)*

Habitats évaporitiques du Silurien du Michigan (Etats-Unis).

Evaporitic habitats of the Michigan Silurian (USA).

PLANCHE 74 HABITAT DE LA HALITE. SA MOBILITE

Fig. 1 Laminites (dolomie-halite). Toute la halite est claire. Les gros cristaux interlamines ont manifestement poussé en repoussant vers le bas et le haut les couches sédimentaires, impliquant leur plasticité rémanente. Avant de penser "remplacement" il faut noter le sentiment que même les couches de boue carbonatées devaient être imprégnées de saumure favorisant la croissance des cristaux.
(Evaporite A_1 basal, forage Pan Am State Kalkaska).

Fig. 2 Bel exemple de la "cristallotopic anhydrite" des auteurs anglo-saxons. L'exemple précédent permet de pressentir que ces cristaux de halite ont poussé individuellement dans une boue laminée imbibée de saumure.
(Même niveau stratigraphique, forage Paul et Trapp).

Fig. 3 Au-dessus de laminites carbonates-anhydrite une boue dolomitique est envahie de cristaux de halite. Vers la droite des figures angulaires permettent de déceler une **pseudomorphose de gypse en halite.**
Comme démontré par Ogniben et Shearman, nous n'avons pas affaire à des "fer de lance" libres, mais à des cristaux primitifs de gypse ayant des inclusions en arête de poisson (jonctions en pointe vers le haut). D'après ses observations à Abu Dhabi et dans le Permien du Castile, Shearman conclut à la poussée de tels cristaux près de l'interface sédiment-air de boues de Sebkhas.
(Début du cycle A_1 près du sommet d'un récif, forage Pan Am Kalkaska).

PLATE 74 THE HABITAT OF HALITE. ITS MOBILITY

Fig. 1 Laminites (dolomite-halite). All the halite is light coloured. The large interlaminar crystals grew, it is obvious, by pushing against upper and lower sedimentary layers, a fact which implies their remanent mouldability.
Before automatically attributing this to a replacement process, it must be realised that, to encourage growth of crystals, even the carbonated mud layers must have been impregnated with brines.
(A_1 basal evaporite, Pan Am_1 well State Kalkaska).

Fig. 2 A good example of "crystallotopic anhydrite".
The previous case suggests that the halite crystals grew individually in a laminated mud, saturated with brine.
(Same stratigraphic layer, Paul and Trapp well).

Fig. 3 Above the carbonate laminites-anhydrite, a dolomitic mud flooded by halite crystals. To the right angular shapes reveal **pseudomorphism of gypsum into halite.**
As Ogniben and Shearman showed, it is not a question of free "swallow-tails" but rather of primitive gypsum crystals including fish bones inclusions (junctions pointing upwards). From his observations in Abu-Dhabi and in the Permian of Castile, Shearman concludes in the growth of such crystals near the sediment-air interface in the Sebkha muds.
(Start of A_1 cycle near the top of the reef, Pan Am well, Kalkaska).

1

2

3

CHAPITRE XI (Planches 75 à 87). *CHAPTER XI (Plates 75 to 87)*

Evaporites d'un bassin continental subsident (Ludien et Stampien de Mormoiron-Pernes, sud-est de la France). Aspects séquentiels du dépôt; faciès primaires et leur évolution diagénétique.

Evaporites in subsident continental basin (Ludian and Stampian of Mormoiron-Pernes, in the South-East of France). Sequential aspects of deposition; primary facies and their diagenetic evolution.

PLANCHE 75 ASPECT SEQUENTIEL DE LA SEDIMENTATION EVAPORITIQUE LUDIENNE DANS LE BASSIN DE MORMOIRON (Vaucluse, sud-est de la France)

Fig. 1 Sur les marges du bassin évaporatoire (Mormoiron, niveau E, fig. 21) la périodicité des apports détritiques détermine une alternance gypse/argiles vertes à la base de la série. Au-dessus, l'aspect cyclique est réglé par une alternance gypse/argiles + carbonates où les matériaux détritiques cèdent la place très progressivement à une paragenèse à dolomite et sépiolite (milieu de la falaise). Les variations minéralogiques et celles des faciès (*cf.* fig. 21 in-texte) traduisent l'évolution du milieu vers une aridité croissante.
Carrière de Mormoiron S.

Fig. 2 Empreinte de mammifère (ongulé didactyle) à la limite argile verte/gypse. La netteté et la profondeur de cette trace indiquent une plasticité importante du matériau. D'autres traces, plus légèrement imprimées, se rencontrent sur des surfaces argileuses à *mud-cracks* (matériau déjà fortement asséché ; Triat et Truc, 1972). Les bancs de gypse renferment également des empreintes mais leur densité la plus importante est toujours concentrée aux limites supérieures des bancs argileux, qui correspondent au retour à des phases plus arides, aussi bien sur les marges qu'au centre du bassin. Ces caractères traduisent à l'évidence une très faible bathymétrie du milieu et des périodes d'assèchement total du bassin.
Carrière de Mormoiron.

Fig. 3 Il existe également, à l'intérieur des bancs de gypse, une alternance très marquée, à une échelle beaucoup plus fine : dolomicrite à sépiolite (niveaux blancs)/gypse primaire (saccharoïde ou en monocristaux lenticulaires) ; dans cet ensemble, des cristaux de diagenèse précoce germent et donnent en surface le faciès "pieds d'alouette".
Carrière de Mazan, centre du bassin, niveau D.

PLATE 75 SEQUENTIAL ASPECT OF THE LUDIAN EVAPORITIC SEDIMENTATION IN THE MORMOIRON BASIN (Vaucluse, Southeastern France)

Fig. 1 On the margins of the evaporatory basin (Mormoiron, Level, E, Fig. 21) the periodicity of the detritic deposits determines a gypsum/green clay alternation at the base of the series. Above this, the cyclic aspect is controlled by a gypsum/clay + carbonates alternation, where the detritic materials gradually give way to a dolomite and sepiolite paragenesis (middle of the cliff). The mineralogic variations, together with those of the facies (*cf.* Fig. 21 in text), reflect the evolution of the environment towards a growing aridity.
Mormoiron S quarry.

Fig. 2 Print of a mammal (onguleted didactyle) at the green clay/gypsum boundary. The depth and clearness of this mark indicates an important plasticity of the material. Other, more lightly printed marks can be found on the argillaceous mud-cracks surfaces (material already strongly dried : Triat and Truc, 1972). The gypsum banks also include imprints but their most important density is always concentrated in the upper limits of the argillaceous banks, which corresponds to a return to more arid phases, both in the edges and in the centre of the basin. These characteristics reflect a very weak bathymetry of the environment and periods of total drying-out of the basin.
Mormoiron quarry.

Fig. 3 In the interior of the gypsum banks, there also exists a very marked alternation on a much finer scale : dolomicrite to sepiolite (white layers)/primary gypsum (saccharoïde or in lenticular monocrystals). In this unity, the crystals of early weathering diagenesis germinate, and give on the surface the "lark's foot" facies.
Mazan quarry, centre of basin, level D.

1
2
3

CHAPITRE XI (Planches 75 à 87). *CHAPTER XI (Plates 75 to 87)*

Evaporites d'un bassin continental subsident (Ludien et Stampien de Mormoiron-Pernes, sud-est de la France). Aspects séquentiels du dépôt; faciès primaires et leur évolution diagénétique.

Evaporites in subsident continental basin (Ludian and Stampian of Mormoiron-Pernes, in the South-East of France). Sequential aspects of deposition; primary facies and their diagenetic evolution.

PLANCHE 76 FACIES PRIMAIRES ET DE DIAGENESE PRECOCE

Fig. 1 Monocristaux à faces planes (dont certains présentent des macles de cristaux à axes parallèles) dans un sédiment dolomicritique à sépiolite. Les nucléations qui aboutissent à ce faciès cristallin sont rares et semblent liées à une syngenèse-diagenèse très précoce dans les premiers centimètres du sédiment.
Carrière de Mazan, centre du bassin, niveau D (fig. 21 in-texte).
Lumière polarisée, × 14.

Fig. 2 Monocristaux lenticulaires, orientés parallèlement à la lamination, dans un sédiment argilo-dolomitique riche en matière organique. La nucléation de tels germes semble se faire dans la masse des eaux mères ; elle est suivie d'une accrétion diagénétique une fois les cristaux déposés sur le fond.
Carrière de Mazan, niveau B, fig. 21 in-texte.
Lumière polarisée, × 14.

Fig. 3 De bas en haut, l'augmentation très nette du nombre de germes traduit vraisemblablement une salinité croissante du milieu et une décroissance de l'évolution diagénétique par accrétion après le dépôt. Même localisation que fig. 20.
Lumière demi-polarisée, × 14.

Fig. 4 Aspect le plus fréquent du microfaciès des gypses primaires : grande densité des germes lenticulaires dont la nucléation dans les eaux mères se poursuit par une diagenèse précoce à l'interface eau - sédiment (ou dans le sédiment) qui provoque un accroissement de taille des cristaux. Même localisation que fig. 22.
Lumière polarisée, × 34,5.

Fig. 5 Nodule d'anhydrite : faciès de diagenèse précoce, très rarement conservé à l'état d'anhydrite, du type "golfe-Persique" : cristallogenèse dans une tranche de sédiments sur les marges du bassin.
Carrière de Mazan, base du niveau C (fig. 21 in-texte).
Lumière polarisée, × 34,5.

PLATE 76 PRIMARY FACIES AND EARLY DIAGENESIS

Fig. 1 Monocrystals with flat sides (certain of which show twinned crystals with parallel axes) in a dolomicritic sediment with sepiolite. The nucleations which lead to these crystalline facies are rare, and seem linked to a very early syngenesis-diagenesis, in the first centimetres of sediment.
Mazan quarry, centre of the basin, level D (Fig. 21 in text).
Polarized light, × 14.

Fig. 2 Lenticular monocrystals, orientated in parallel towards laminations in an argillaceous-dolomitic sediment, rich in organic matter. The nucleation of such germs seems to occur in the mass of the source-waters ; it is followed by a diagenetic accretion once the crystals are deposited on the bottom.
Mazan quarry, Level B (Fig. 21 in text).
Polarized light, × 14.

Fig. 3 From top to bottom, the very clear increase in the number of germs probably represents a growing salinity in the environment, and a decrease in the diagenetic evolution by accretion, after deposition. Same location as Fig. 2.
1/2 polarized light, × 14.

Fig. 4 The most common aspect of the primary gypsum microfacies : high density of lenticular germs. Their nucleation in the source-waters being continued by an early diagenesis at the water-sediment (or in the sediment) interface, which causes an increase in crystal size. Same location as Fig. 2.
Polarized light, × 34,5.

Fig. 5 Anhydrite nodule ; facies of early diagenesis, very rarely conserved in the anhydrite state, of the "Persian Gulf" type: crystallogenesis in a slice of sediments on the basin's margins. Mazan quarry, base of Level C, (Fig. 21 in text).
Polarized light, × 34,5.

1

2

3

4

5

CHAPITRE XI (Planches 75 à 87). *CHAPTER XI (Plates 75 to 87)*

Evaporites d'un bassin continental subsident (Ludien et Stampien de Mormoiron-Pernes, sud-est de la France). Aspects séquentiels du dépôt; faciès primaires et leur évolution diagénétique.

Evaporites in subsident continental basin (Ludian and Stampian of Mormoiron-Pernes, in the South-East of France). Sequential aspects of deposition; primary facies and their diagenetic evolution.

PLANCHE 77 FACIES PRIMAIRES ET DE DIAGENESE PRECOCE : BIOTURBATIONS

Fig. 1 Section d'un terrier vertical ; la différence de taille et de faciès entre les cristaux qui remplissent le terrier et ceux qui constituent le sédiment encaissant traduit des variations rapides de salinité ; les microcristaux saccharoïdes (en remplissage sur cette figure) ont germé dans des eaux mères plus riches en sulfates que celles ayant engendré les cristaux lenticulaires de l'encaissant. Carrière de Mazan, centre du bassin, niveau B (Fig. 21 in-texte).
Lumière polarisée, × 14.

Fig. 2 Autre section de même nature, mais phénomène inverse de remplissage et de variation dans la salure des eaux mères. Même localisation.
Lumière polarisée, × 14.

Fig. 3 Section d'un terrier parallèle à la sédimentation. Les bioturbants provoquent une réorganisation grossièrement circulaire des cristaux antérieurement formés. Même localisation.
Lumière polarisée, × 14.

Fig. 4 et 5 L'activité des fouisseurs, peu fréquemment citée dans les dépôts évaporitiques, provoque une désorganisation très poussée du sédiment, en particulier lorsque les traces de fouissage sont aussi bien verticales qu'obliques (sommet de la fig. 5).

L'ensemble de ces traces témoigne d'un milieu périodiquement peu salé, abritant au moins certaines formes de vie (essentiellement des larves d'insectes). Même localisation.
Lumière polarisée, × 14.

PLATE 77 PRIMARY FACIES AND EARLY DIAGENESIS : BIOTURBATIONS

Fig. 1 Section of a vertical burrowing ; the difference in size and facies between the crystals inside the burrowing and those which form the sediment surrounding it, reflects the rapid salinity variations ; the saccharoïde microcrystals (in "filling", on the figure) have germinated in source waters more rich in sulphates than those which have produced the lenticular crystals of the area around it. Mazan quarry, level B (Fig. 21 in text).
Polarized light, × 14.

Fig. 2 Another section of the same kind, but with the opposite phenomenon of filling and variation in the salinity of the source-waters. Same location.
Polarized light, × 14.

Fig. 3 Section of a burrowing parallel to the sedimentation. The burrowers cause a roughly circular reorganization of the crystals formed previously. Same localization.
Polarized light, × 14.

Figs. 4 and 5 The activity of burrowers, very rarely mentioned in evaporitic deposits, causes a considerable sedimental disorder, particularly when the burrowings are both vertical and oblique (top of Fig. 5). All these indications seem to represent a periodically hyposaline environment, containing at least some forms of life (mainly insect larva). Same location.
Polarized light, × 14.

1

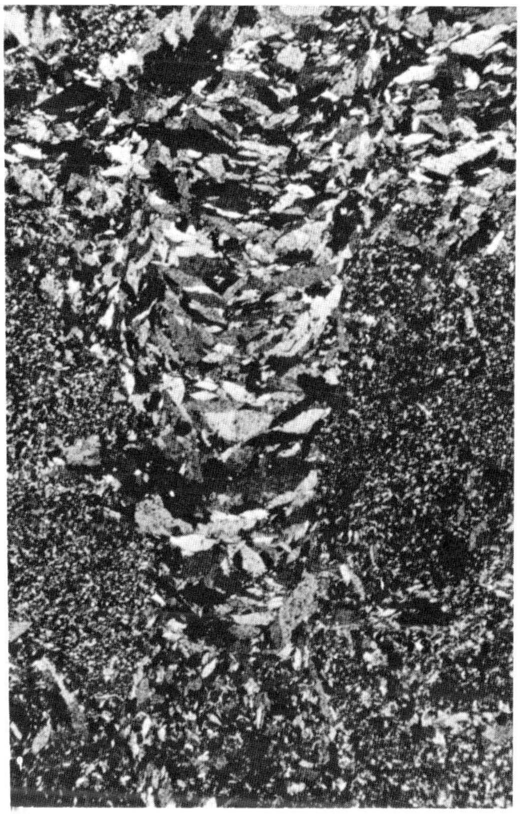

2

3

4

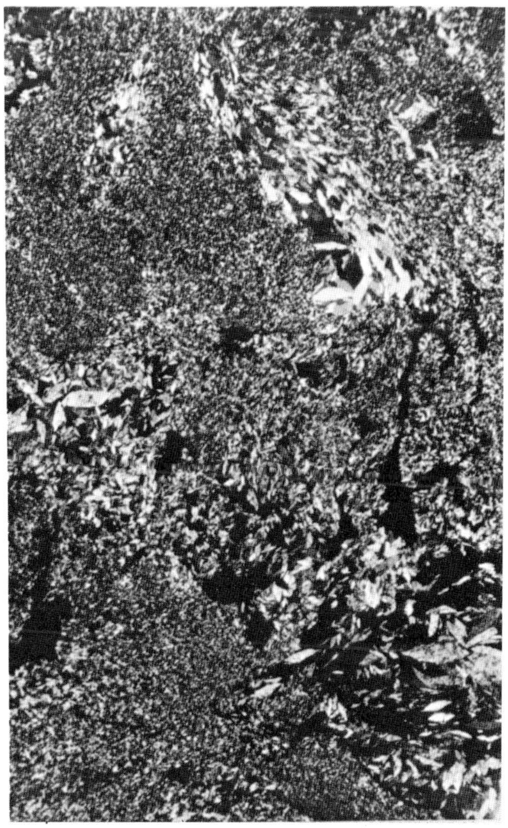

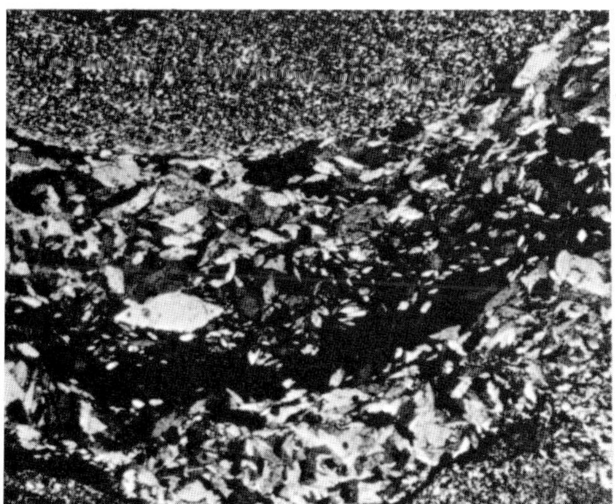

5

CHAPITRE XI (Planches 75 à 87). *CHAPTER XI (Plates 75 to 87)*

Evaporites d'un bassin continental subsident (Ludien et Stampien de Mormoiron-Pernes, sud-est de la France). Aspects séquentiels du dépôt; faciès primaires et leur évolution diagénétique.

Evaporites in subsident continental basin (Ludian and Stampian of Mormoiron-Pernes, in the South-East of France). Sequential aspects of deposition; primary facies and their diagenetic evolution.

PLANCHE 78 DIAGENESE PRECOCE : CROISSANCE DES CRISTAUX ET DEFORMATIONS SEDIMENTAIRES

Fig. 1 Monocristaux diagénétiques de grande taille. De telles cristallogenèses aboutissent au faciès "pieds d'alouette". Ces cristaux perforent certains lits de gypse et provoquent une déformation des niveaux argilo-dolomitiques sus-jacents. Carrière de Mazan, centre du bassin, niveau D (fig. 21 in-texte).
Lumière naturelle, x 14.

Fig. 2 Au cours de leur croissance, ces monocristaux peuvent atteindre l'interface eau-sédiment et induire la formation de micro-cuvettes qui subissent un remplissage préférentiel. Noter l'alternance gypse saccharoïde à cristaux sub-coalescents/gypse en cristaux lenticulaires dispersés dans l'argile (salure à variations cycliques des eaux mères). Carrière de Mazan, même niveau.
Lumière naturelle, x 14.

Fig. 3 Déformations des laminites argileuses à hydrocarbures par l'accrétion de germes lenticulaires. Carrière de Mazan, niveau B (fig. 21 in-texte).
Lumière naturelle, x 34,5.

Fig. 4 Des rosettes de "cristaux satellites" sont fréquemment accolées aux monocristaux diagénétiques de grande taille. Même localisation que fig. 1.
Lumière polarisée, x 14.

Fig. 5 Déformations de sédiments argileux riches en matière organique sous l'effet du même type de phénomène ; les lits de gypse montrent plusieurs monocristaux à faces planes ayant eux-mêmes subi une accrétion par diagenèse précoce. Noter les inclusions de matière organique et d'argile dans les grands monocristaux. Même localisation que fig. 1.
Lumière naturelle, x 14.

PLATE 78 EARLY DIAGENESIS : GROWTH OF CRYSTALS AND SEDIMENTARY DEFORMATIONS

Fig. 1 Large size diagenetic monocrystals. Such crystallogeneses result in "larks foot" facies. These crystals perforate certain gypsum beds, and cause a deformation of the over-lying argilo-dolomitic layers. Mazan quarry, center of the basin, level D (Fig. 21 in text).
Normal light, x 14.

Fig. 2 During their growth, these monocrystals may reach the water-sediment interface, and induce the formation of micro-basins, which undergo preferential filling. Note the alternation between saccharoïde gypsum with subcoalescent crystals, and lenticular gypsum dispersed in the clay (varied cyclic salinity of the source-waters). Mazan quarry, same level.
Normal light, x 14.

Fig. 3 Deformations of the hydrocarbon argillaceous laminites by lenticular germ accretion. Mazan quarry, level B (Fig. 21 in text).
Normal light, x 34,5.

Fig. 4 Spherules of "satellite crystals" are often attached to large size monocrystals. Same location as Fig. 1.
Normal light, x 14.

Fig. 5 Deformations of argillaceous sediments rich in organic matter, under the effect of the same type of phenomenon : the gypsum beds show several monocrystals with flat sides, which have themselves undergone accretion by early diagenesis. Note the inclusions of organic matter and clay, in the large monocrystals. Same location as Fig. 1.
Normal light, x 14.

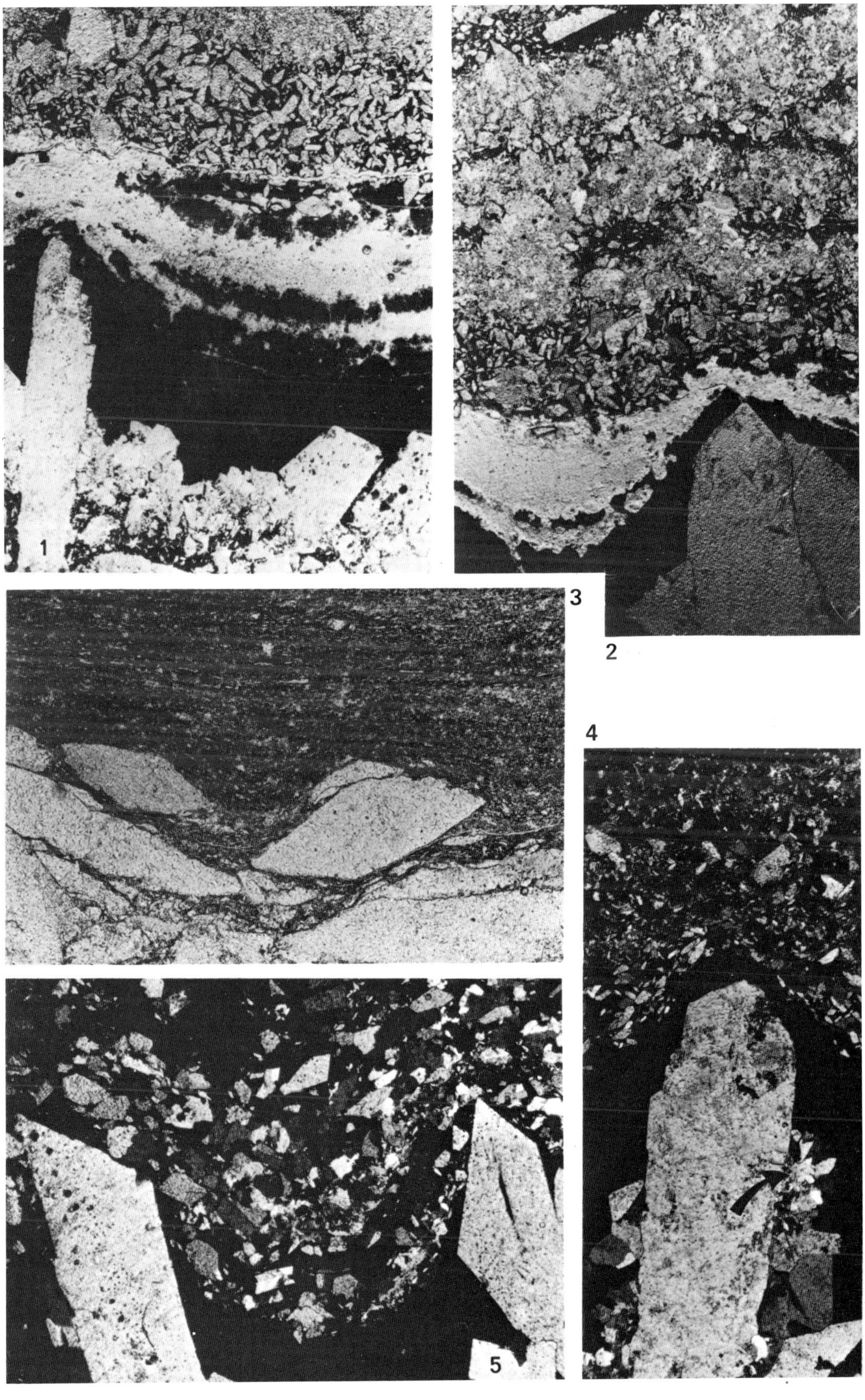

CHAPITRE XI (Planches 75 à 87). *CHAPTER XI (Plates 75 to 87)*

Evaporites d'un bassin continental subsident (Ludien et Stampien de Mormoiron-Pernes, sud-est de la France). Aspects séquentiels du dépôt; faciès primaires et leur évolution diagénétique.

Evaporites in subsident continental basin (Ludian and Stampian of Mormoiron-Pernes, in the South-East of France). Sequential aspects of deposition; primary facies and their diagenetic evolution.

PLANCHE 79 DIAGENESE PRECOCE : REMPLACEMENTS ET INCLUSIONS

Fig. 1 De nombreuses surfaces de bancs montrent l'existence d'un voile algaire rétracté, intégralement épigénisé par du gypse (le marteau donne l'échelle). Carrière de Mazan, limite niveaux B et C (fig. 21 in-texte).

Fig. 2 En lame mince, de tels niveaux présentent une grande densité de filaments algaires, emprisonnés dans des monocristaux lenticulaires de grande taille (diagenèse précoce). Carrière de Mazan, niveau D (fig. 21 in-texte).
Lumière demi-polarisée, × 34,5.

Fig. 3 Même phénomène : ici, le gypse remplace le ciment argilo-dolomicritique et cristallise en grandes plages monocristallines mal délimitées. Même localisation.
Lumière demi-polarisée, × 34,5.

Fig. 4 Détail des filaments : la matrice organique constituant la gaine est sporadiquement conservée ; seul un remplissage argilo-carbonaté permet la plupart du temps la mise en évidence de telles structures. Des filaments semblables, appartenant au genre *Microcoleus*, ont été signalés par C. Schreiber (1973) dans les évaporites messiniennes de Méditerranée. Même localisation.
Lumière naturelle, × 140.

Fig. 5 Detail de la structure (fig. 1) à la surface des bancs.

PLATE 79 EARLY DIAGENESIS : REPLACEMENTS AND INCLUSIONS

Fig. 1 Numerous surfaces of banks show the existence of a retracted algae film integrally epigenized by gypsum (the hammer gives the scale). Mazan quarry, levels B and C (Fig. 21 in text).

Fig. 2 In this section, these levels show a high density of algeous filaments caught in the large-sized lenticular monocrystals (early diagenesis). Mazan quarry, level D (Fig. 21 in text).
1/2 polarized light, × 34,5.

Fig. 3 Same phenomenon here, the gypsum replaces the argilo-dolomicritic cement and crystallizes into large, badly defined, monocrystalline beaches. Same location.
1/2 polarized light, × 34,5.

Fig. 4 Detail of the filaments : the organic matrix forming the casing is sporadically conserved ; only the argilo-carbonated filling generally allows these structures to be seen. Similar filaments, of the *Microcoleus* group, have been pointed out by C. Schreiber (1973) in the Messinian evaporites of the Mediterranean. Same location.
Normal light, × 140.

Fig. 5 Detail of the structure (Fig. 1) at the surface of the banks.

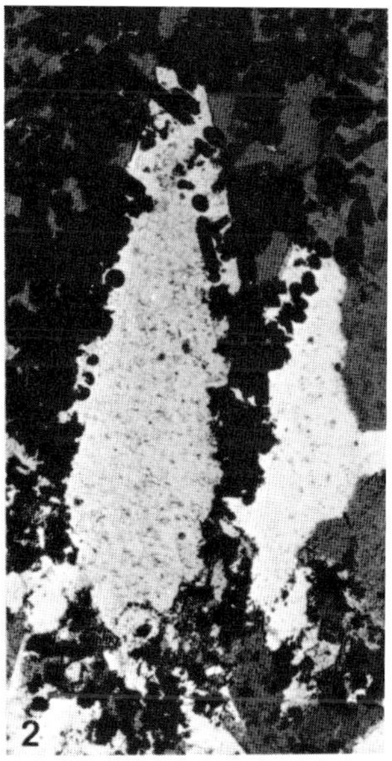

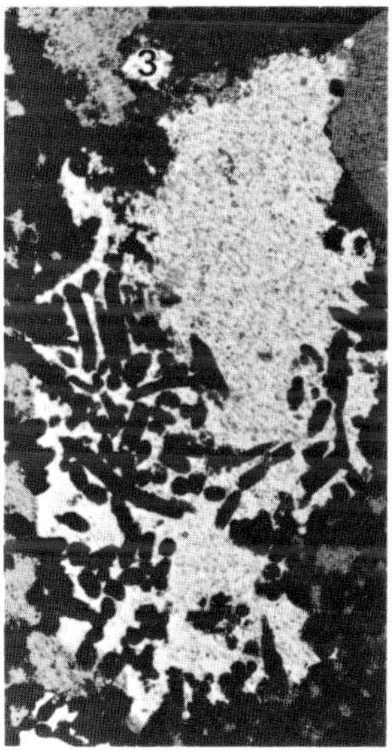

4

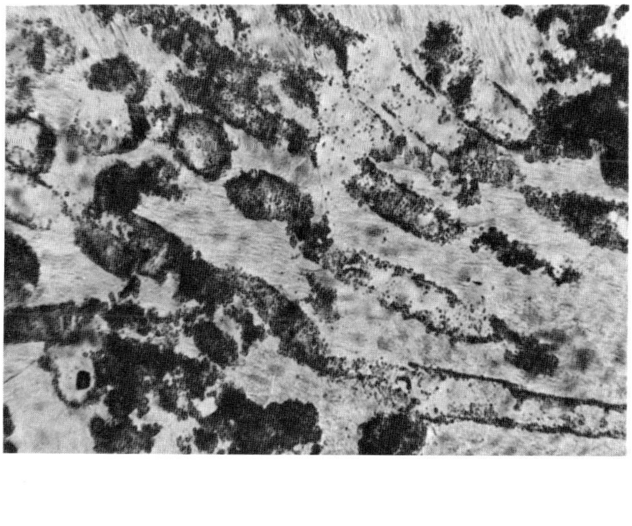

5

CHAPITRE XI (Planches 75 à 87). *CHAPTER XI (Plates 75 to 87)*

Evaporites d'un bassin continental subsident (Ludien et Stampien de Mormoiron-Pernes, sud-est de la France). Aspects séquentiels du dépôt; faciès primaires et leur évolution diagénétique.

Evaporites in subsident continental basin (Ludian and Stampian of Mormoiron-Pernes, in the South-East of France). Sequential aspects of deposition; primary facies and their diagenetic evolution.

PLANCHE 80 DIAGENESE PRECOCE : ACCRETION POLYPHASEE-CORROSION

Fig. 1 Des plages monocristallines de diagenèse précoce présentent fréquemment de nombreuses inclusions : cristaux de gypse lenticulaires, baguettes de gypse en substitution sur de l'anhydrite (flèche), rhomboèdres de dolosparite dissous et remplis de sédiments argileux (en haut à droite), plages argilo-dolomicritiques ; noter, au centre de l'image, un monocristal de gypse en partie corrodé (vers le bas) montrant lui-même des inclusions. Ce type de microfaciès traduit la précarité de l'équilibre qui existe entre le domaine des cristallogenèses déjà élaborées et celui des solutions qui circulent dans les premiers centimètres du dépôt, au cours de la diagenèse précoce. Ce sont des variations dans la mobilité et la concentration des eaux mères du bassin évaporatoire qui assurent le contrôle de ces phénomènes dont le résultat apparaît comme une accrétion très polyphasée s'effectuant *in fine* au bénéfice de la phase gypseuse. Carrière de Mazan, niveau B (fig. 21 in-texte).
Lumière polarisée, x 34,5.

Fig. 2 Mêmes phénomènes ; inclusions de gypse selon des plans préférentiels du réseau cristallin (flèches droites) et d'anhydrite (flèche courbe). Vers le haut, inclusions de cristaux de gypse en substitution sur de l'anhydrite. Noter les déformations du sédiment produites par ce type d'accrétion. Même localisation.
Lumière polarisée, x 34,5.

Fig. 3 et 4 Certains niveaux riches en grands cristaux de diagenèse précoce souvent maclés présentent des figures de corrosion lorsqu'une rupture dans la sédimentogenèse provoque une phase d'érosion, même de très faible envergure et sur des surfaces restreintes. Les têtes de cristaux sont alors arasées, plus vraisemblablement par dissolution que par un effet mécanique. Une génération de cristaux primaires, saccharoïdes ou lenticulaires, suit la phase d'érosion et traduit un retour à la présence dans le bassin d'eaux mères en voie de saturation (vis-à-vis du gypse). Ce processus est cyclique : les fig. 3 et 4 montrent nettement trois générations de monocristaux arasés (la flèche indique un point de silicification). Carrière de Mazan, niveau D (fig. 21 in-texte).
Lumière polarisée, x 14.

PLATE 80 EARLY DIAGENESIS : POLYPHASEAN ACCRETION-CORROSION

Fig. 1 Monocrystalline beaches of early diagenesis frequently present numerous inclusions : lenticular gypsum crystals, laths of gypsum in substitution on the anhydrite (arrow), rhombohedrons of dolosparite below, and filled with argillaceous sediments (top right), argilo-dolomicritic beaches ; note in the centre of the picture a gypsum monocrystal partly corroded (towards the bottom), itself showing inclusions. This type of microfacies demonstrates the precarious balance which exists as regards the already elaborated crystallogenesis, and the solutions which circulate in the first centimetres of deposition during the early diagenesis. It is the variations in the mobility and concentration of the source-water of the evaporatory basin which control these phenomena. The result appears as a very polyphased accretion made at the end to the benefit of the gypseous phase. Mazan quarry, level B (Fig. 21 in text).
Polarized light, x 34,5.

Fig. 2 Same phenomena : inclusions of gypsum following the presential planes of the crystalline network (straight arrows) and of anhydrite (curved arrow). Towards the top inclusions of gypsum crystals in substitution on anhydrite. Note the deformations of sediment produced by this type of accretion. Same location.
Polarized light, x 34,5.

Figs. 3 and 4 Certain levels rich in large crystals of early diagenesis, often twinned, present figures of corrosion when a rupture in sedimentogenesis causes a phase of erosion, even on a very small scale on limited surfaces. The crystal heads are then struck off, more probably by dissolution, than by any mechanical effect. A generation of primary crystals, saccharoïde or lenticular, follows the erosion phase, and reflects a return to the presence in the basin of source water becoming saturated (as regards the gypsum). This is a cyclic process : Figs. 3 and 4 clearly show the three generations of struckoff monocrystals (the arrow indicated a point of silicification). Mazan quarry Level D (Fig. 21 in text).
Polarized light, x 14.

1

2

3

4

CHAPITRE XI (Planches 75 à 87). *CHAPTER XI (Plates 75 to 87)*

Evaporites d'un bassin continental subsident (Ludien et Stampien de Mormoiron-Pernes, sud-est de la France). Aspects séquentiels du dépôt; faciès primaires et leur évolution diagénétique.

Evaporites in subsident continental basin (Ludian and Stampian of Mormoiron-Pernes, in the South-East of France). Sequential aspects of deposition; primary facies and their diagenetic evolution.

PLANCHE 81 FACIES DIAGENETIQUES : TRANSFORMATIONS ANHYDRITE EN GYPSE

Fig. 1 A la base, matrice argilo-carbonatée renfermant de petits cristaux primaires ; au sommet, anciens nodules anhydritiques totalement transformés en gypse : la limite entre ces deux faciès est marquée par un lit de gypse fibreux. Ce phénomène présente une remarquable constance dans les dépôts examinés et traduit l'existence de mouvements différentiels entre les lits nodulaires en voie de transformation (augmentation de volume) et la masse inerte des sédiments primaires. Carrière de Mazan, niveau C (fig. 21 in-texte)
Lumière polarisée, x 14.

Fig. 2 Détail du même phénomène ; noter l'aspect engrené très particulier des cristaux de gypse provenant de la transformation totale anhydrite/gypse. Même localisation.
Lumière polarisée, x 14.

Fig. 3 Anciens nodules anhydritiques, partiellement transformés en gypse, dans une matrice argilo-carbonatée. De tels nodules ont pris naissance, selon toute vraisemblance, dans des sédiments exondés, au niveau de la frange capillaire, selon le modèle décrit par les auteurs sur la Trucial Coast du golfe Persique (Shearman, 1963, 1966). Carrière de Mazan, niveau C (fig. 21 in-texte).
Lumière polarisée, x 14.

Fig. 4 Restes d'anhydrite en "lattes", fragmentés et corrodés sur place, dans le même type de nodules. Même localisation.
Lumière polarisée, x 14.

Fig. 5 Détail de la figure précédente ; les lattes d'anhydrite résiduelle possèdent une disposition parallèle ou radiaire. Même localisation.
Lumière polarisée, x 34,5.

PLATE 81 DIAGENETIC FACIES : ANHYDRITE INTO GYPSUM

Fig. 1 At the base, an argilo-carbonate matrix enclosing small primary crystals : at the top, former anhydritic nodules completely transformed into gypsum. The boundary between these two facies is marked by a bed of fibrous gypsum. This phenomenon shows a remarkable persistance in the deposits examined, and respects the existence of differentiel movements between the nodular beds during transformation (increase in volume), and the inert mass of primary sediments. Mazan quarry, level C (Fig. 21 in text).
Polarized light, x 14.

Fig. 2 Detail of the same phenomenon : note the very particular meshed aspect of the gypsum crystals coming from the total anhydrite/gypsum transformation. Same location.
Polarized light, x 14.

Fig. 3 Former anhydritic nodules, partially transformed into gypsum, in an argilo-carbonate matrix : these nodules very probably occured in emerged sediments, on the capillary border level, following the model described, on the Trucial Coast of the Persian Gulf (Shearman, 1963-1966). Mazan quarry, level C (Fig. 21 in text)
Polarized light, x 14.

Fig. 4 Anhydrite remains in "laths", fragmented and corroded in the area, in the same type of nodules. Same location.
Polarized light, x 14.

Fig. 5 Detail of the previous photo : the laths of residual anhydrite possess a parallel or radial arrangement. Same location.
Polarized light, x 34,5.

1
2
3
4
5

CHAPITRE XI (Planches 75 à 87). *CHAPTER XI (Plates 75 to 87)*

Evaporites d'un bassin continental subsident (Ludien et Stampien de Mormoiron-Pernes, sud-est de la France). Aspects séquentiels du dépôt; faciès primaires et leur évolution diagénétique.

Evaporites in subsident continental basin (Ludian and Stampian of Mormoiron-Pernes, in the South-East of France). Sequential aspects of deposition; primary facies and their diagenetic evolution.

PLANCHE 82 FACIES DIAGENETIQUES : SUBSTITUTIONS DE GYPSE PAR L'ANHYDRITE.

Fig. 1 à 4 Monocristaux de gypse primaire, totalement remplacés par de l'anhydrite. L'absence de déformation des structures primaires implique une dissolution préalable du gypse et la cristallisation d'anhydrite à partir de solutions sulfatées-calciques ayant imbibé le sédiment sous des conditions particulières de température (*cf.* Jauzein, 1974). De telles diagenèses sont typiques des niveaux profonds du bassin de Mormoiron, caractérisés par une sédimentation argileuse riche en matière organique (laminites à hydrocarbures), à niveaux de gypse primaire peu abondants.

Dans certains cas, (fig. 2), une évolution postérieure fait apparaître une dentelle de gypse sur les bords des cristaux.

Carrière de Mazan, niveau A, fig. 21 in-texte.

Lumière polarisée x 34,5.

PLATE 82 DIAGENETIC FACIES : GYPSUM ANHYDRITE SUBSTITUTIONS

Figs. 1 to 4 Primary gypsum monocrystals, totally replaced by anhydrite. The lack of deformation of the primary structures implies a previously existing gypsum solution, and the crystallization of anhydrite based on sulfate-calcic solutions which had saturated the sediment under particular temperature conditions (*cf.* Jauzein, 1974). Such diageneses are typical of the deep layers of the Mormoiron basin, characterized by an argillaceous sedimentation rich in organic matter (hydrocarbon laminates), at levels where primary gypsum is fairly uncommon.

In certain cases (Fig. 2) a previous evolution causes a lace-work of gypsum to appear on the edges of the crystals.

Mazan quarry, level A, Fig. 21 in text

Polarized light, x 34,5.

1

2

3

4

CHAPITRE XI (Planches 75 à 87). *CHAPTER XI (Plates 75 to 87)*

Evaporites d'un bassin continental subsident (Ludien et Stampien de Mormoiron-Pernes, sud-est de la France). Aspects séquentiels du dépôt; faciès primaires et leur évolution diagénétique.

Evaporites in subsident continental basin (Ludian and Stampian of Mormoiron-Pernes, in the South-East of France). Sequential aspects of deposition; primary facies and their diagenetic evolution.

PLANCHE 83 FACIES DIAGENETIQUES : SUBSTITUTIONS DU GYPSE PAR LA SILICE

Fig. 1 à 3 Les substitutions de silice sur le gypse affectent essentiellement des niveaux riches en monocristaux lenticulaires primaires, ou bien les grands cristaux de diagenèse précoce. Elles se traduisent par les formes : calcédonite, quartzine et lutécite.
Fig. 1 et 2, carrière de Mazan, niveau B, fig. 21 in-texte ; fig. 3, niveau D.
Lumière polarisée, x 34,5.

Fig. 4 Silicification précoce : lit de monocristaux diagénétiques (flèche) remplacés par de la lutécite, dans un sédiment argilo-micritique. Seul le niveau à monocristaux est remplacé.
Stampien de Pernes, près Mormoiron.
x 1.

Fig. 5 Détail de la figure précédente : aspect typique de la lutécite remplaçant la totalité d'un monocristal lenticulaire.
Même localisation.
Lumière polarisée, x 46.

PLATE 83 DIAGENETIC FACIES : GYPSUM SUBSTITUTION BY SILICA

Figs. 1 to 3 The silica substitutions on the gypsum mainly affect the layers rich in lenticular monocrystals, or else the large crystals of early diagenesis. They are reflected in the forms : calcedony, quartzine and lutecite. Figs. 1 and 2, Mazan quarry, level B, fig. 21 in text ; Fig. 3, level D
Polarized light, x 34,5.

Fig. 4 Early silicification : bed of diagenetic monocrystals (arrow), replaced by lutecite, in an argilo-micritic sediment. Only the monocrystals level is replaced.
Stampian of Pernes, near Mormoiron.
x 1.

Fig. 5 Detail of the preceding photo : typical aspect of lutecite replacing the whole of a lenticular monocrystal.
Same location
Polarized light, x 46.

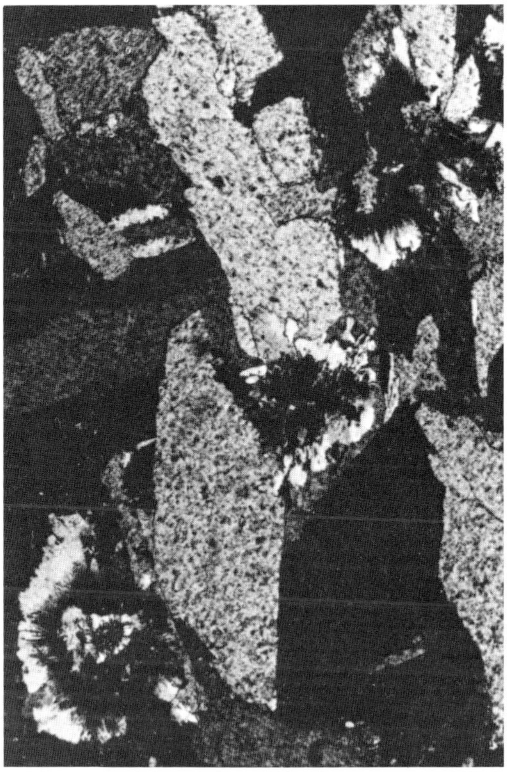

1

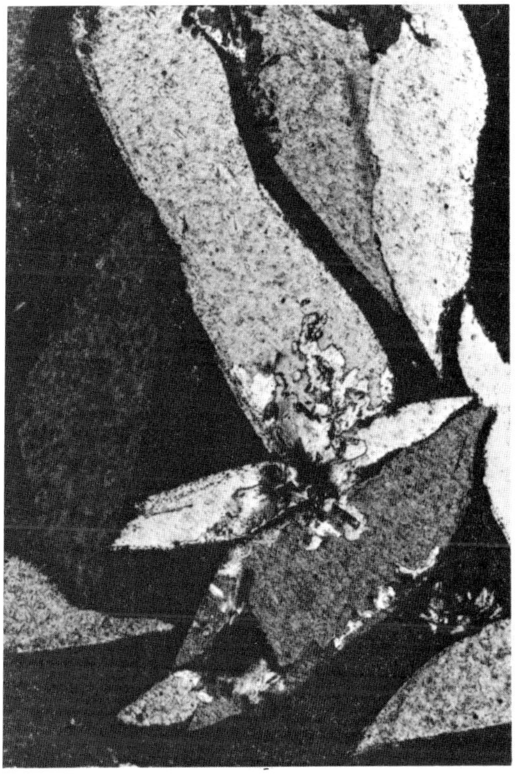

2

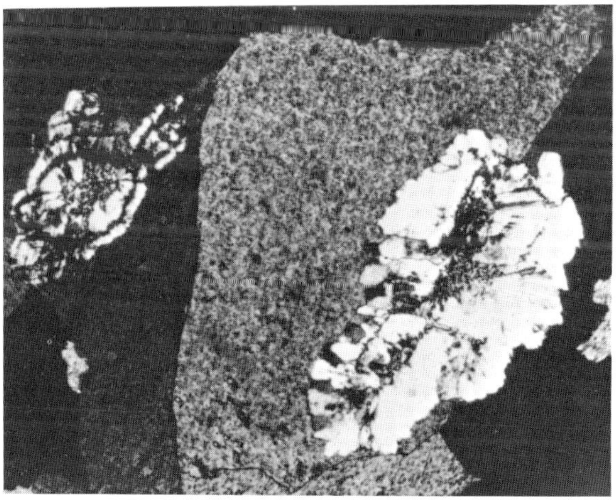

3

4

5

Evaporites d'un bassin continental subsident (Ludien et Stampien de Mormoiron-Pernes, sud-est de la France). Aspects séquentiels du dépôt; faciès primaires et leur évolution diagénétique.

Evaporites in subsident continental basin (Ludian and Stampian of Mormoiron-Pernes, in the South-East of France). Sequential aspects of deposition; primary facies and their diagenetic evolution.

PLANCHE 84 FACIES DIAGENETIQUES : SUBSTITUTIONS DU GYPSE PAR L'ANHYDRITE ET TRANSFORMATIONS EN GYPSE SECONDAIRE

Phénomènes affectant les dépôts des marges du cirque évaporatoire.

Fig. 1 Alternances argiles vertes (détritiques)-gypse sur les marges du bassin ludien de Mormoiron. Mormoiron, "Les Limons", niveau E, fig. 21 in-texte.

Fig. 2 Faciès des bancs de gypse du même secteur. Les laminites argilo-carbonatées (dolomicrites à sépiolite) sont partiellement remplacées au cours de phénomènes de gypsification liés aux mouvements des nappes sur les marges du bassin.

PLATE 84 DIAGENETIC FACIES : GYPSUM SUBSTITUTION BY ANHYDRITE AND SECONDARY GYPSUM

Phenomena affecting the deposits on the edges of the evaporatory basin.

Fig. 1 Green clay (detritic)/gypsum alternations on the edges of the Mormoiron Ludian basin. Mormoiron, "Les Limons", level E, Fig. 21 in text.

Fig. 2 Facies of gypsum banks in the same sector. The argilo-carbonate laminites (sepiolite dolomicrites) are partially replaced during gypsification phenomena, linked to the movements of the nappes on the edges of the basin.

1

2

CHAPITRE XI (Planches 75 à 87). *CHAPTER XI (Plates 75 to 87)*

Evaporites d'un bassin continental subsident (Ludien et Stampien de Mormoiron-Pernes, sud-est de la France). Aspects séquentiels du dépôt; faciès primaires et leur évolution diagénétique.

Evaporites in subsident continental basin (Ludian and Stampian of Mormoiron-Pernes, in the South-East of France). Sequential aspects of deposition; primary facies and their diagenetic evolution.

PLANCHE 85 FACIES DIAGENETIQUES : SUBSTITUTIONS DU GYPSE PAR L'ANHYDRITE ET TRANSFORMATION EN GYPSE SECONDAIRE

Phénomènes affectant les dépôts des marges du cirque évaporatoire.

Fig. 1 Section correspondant à la planche 84, fig. 2 : même phénomène. Le trait noir vaut 3 cm. Même localisation.

Fig. 2 Certaines lames réalisées dans ces gypses montrent quelquefois des sections de cristaux lenticulaires aux contours déformés, tandis que des filonnets de gypse fibreux parcourent le sédiment. Même localisation
Lumière naturelle, x 34,5.

Fig. 3 Même lame, en lumière polarisée : les formes qui paraissent monocristallines en lumière naturelle sont en réalité occupées par un gypse polycristallin présentant des germes très engrenés. Même localisation.
Lumière polarisée, x 34,5.

Fig. 4 Détail de la fig. 3.
Lumière polarisée, x 140.

De tels faciès apparaissent comme un produit de substitution gypse-anhydrite-gypse lié aux mouvements des nappes qui baignent les sédiments marginaux, sous le contrôle des variations de niveau dans le bassin évaporatoire.

PLATE 85 DIAGENETIC FACIES : GYPSUM SUBSTITUTION BY ANHYDRITE AND SECONDARY GYPSUM

Phenomena affecting the deposits on the edges of the evaporatory basin.

Fig. 1 Section corresponding to Plate 84, Fig. 2 ; same phenomenon. The black line represents 3 cm. Same location.

Fig. 2 Certain thin sections carried out in these gypsums sometimes show sections of lenticular crystals with deformed contours, whilst veins of fibrous gypsum cross the sediment. Same location.
Natural light, x 34,5.

Fig. 3 Same thin section, in polarized light : the forms which appear monocrystalline in natural light are in reality occupied by polycristalline gypsum showing very meshed germs. Same location.
Polarized light, x 34,5.

Fig. 4 Detail of Fig. 3.
Polarized light, x 140.

Such facies seem to be a product of substitution gypsum-anhydrite-gypsum linked to movements of the nappes which bathe the marginal sediments controlled by the variations of level in the evaporatory basin.

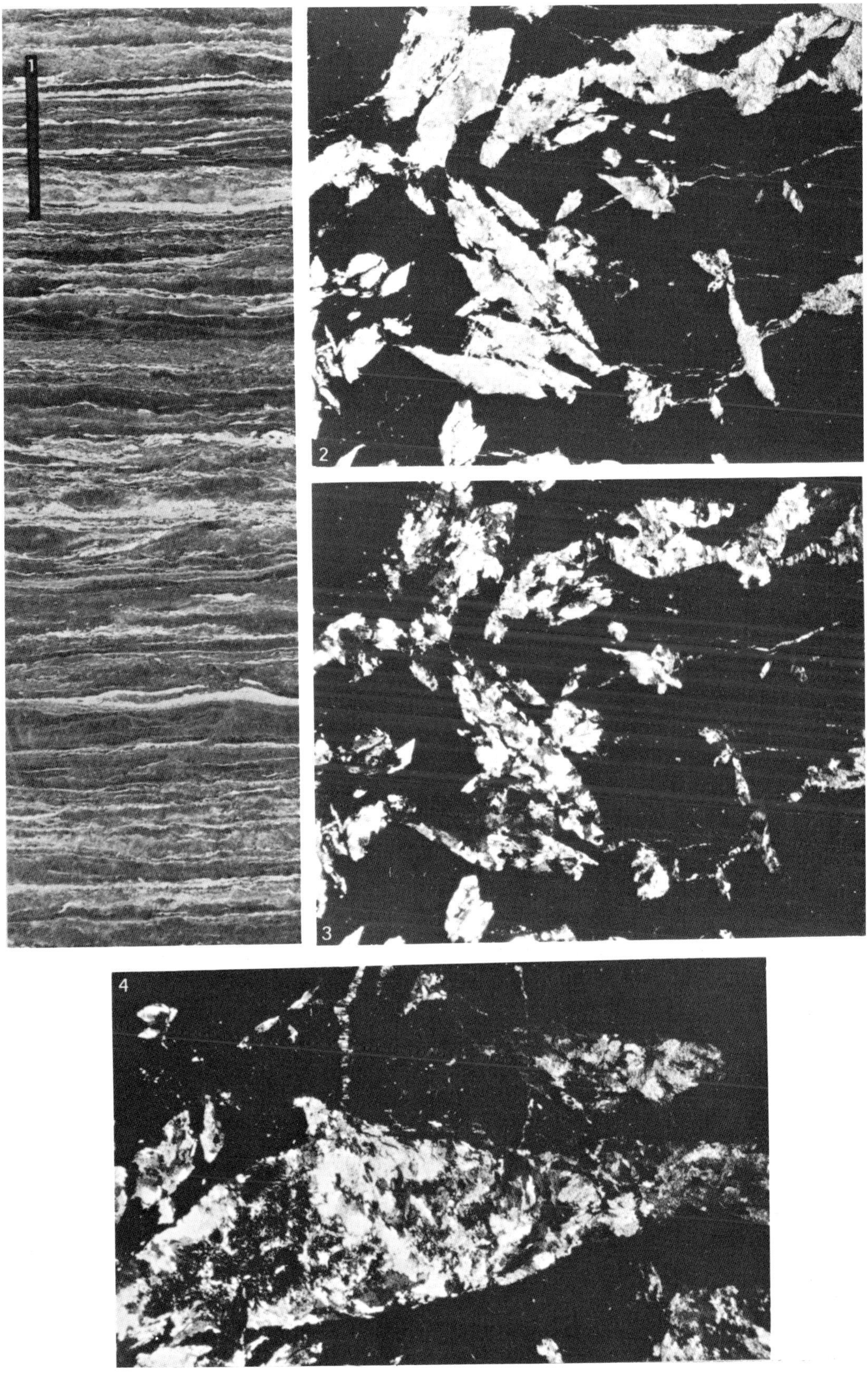
1
2
3
4

CHAPITRE XI (Planches 75 à 87). *CHAPTER XI (Plates 75 to 87)*

Evaporites d'un bassin continental subsident (Ludien et Stampien de Mormoiron-Pernes, sud-est de la France). Aspects séquentiels du dépôt; faciès primaires et leur évolution diagénétique.

Evaporites in subsident continental basin (Ludian and Stampian of Mormoiron-Pernes, in the South-East of France). Sequential aspects of deposition; primary facies and their diagenetic evolution.

PLANCHE 86 FACIES DIAGENETIQUES : SUBSTITUTIONS DE CARBONATES PAR LE GYPSE

Fig. 1 Les évaporites du Stampien de Pernes, près Mormoiron (*cf.* fig. 22 in-texte) dans leur complexe carbonaté (calcitique, avec très rares niveaux dolomitiques). Les bancs gypseux situés au milieu de la falaise, présentent une allure en "oreillers". Ces bancs résultent d'un remplacement de dépôts carbonatés (antérieurement constitués) par un gypse microcristallin. Ces "gypsifications" se produisent sous l'influence d'eaux phréatiques, en provenance des eaux mères du bassin évaporatoire, circulant dans les points bas de la topographie sur les marges du bassin. Quelques niveaux à monocristaux lenticulaires de diagenèse précoce montrent que la surface du bassin augmentait périodiquement et recouvrait amplement les secteurs marginaux. Hauteur de la falaise : 10 m environ.

Fig. 2 Aspect macroscopique des mêmes gypses. Il existe une succession cyclique dans les sédiments carbonatés initialement déposés : laminites micritiques à ostracodes/micrite à oncoïdes, en bancs de 0,50 à 1 m. La figure 2 montre une séquence de laminites micritiques partiellement remplacées par du gypse, et interrompues (milieu et sommet de la figure) par deux niveaux de gypse en grands cristaux lenticulaires, eux-mêmes substitués en lutécite. Carrière de La Roque-sur-Pernes.

Fig. 3 Section d'une séquence de micrite à oncoïdes dont le ciment micritique est presque totalement remplacé par du gypse microcristallin, de faciès albâtre. Même localisation ; chaque subdivision de l'échelle vaut 1 mm.

Fig. 4 En lame mince, aspect des micrites à oncoïdes dans les rares niveaux où le remplacement est inexistant. Ces oncoïdes possèdent une faible cortication ; le nucleus est la plupart du temps un pellet. Même localisation.
Lumière naturelle, × 34,5.

PLATE 86 DIAGENETIC FACIES : CARBONATE SUBSTITUTION BY GYPSUM

Fig. 1 The evaporites of the Pernes Stampian, near Mormoiron (*cf.* Fig. 22 in text) in their carbonate complex (calcitic, with a few very rare dolomitic layers). The gypseous banks, situated in the middle of the cliff, give a "pillow-like" effect. These banks result from a replacement of the carbonate deposits (formed previously), by microcrystalline gypsum. The "gypsification" occurs under the influence of phreatic waters, coming from source-waters of the evaporatory basin, circulating in the low points of topography on the margins of the basin. A few layers with lenticular monocrystals of early weathering diagenesis show that the surface of the basin rose periodically and easily covered the marginal sectors. Height of the cliff : approximately 10 m.

Fig. 2 Macroscopical aspect of the same gypsums. A cyclic succession exists in the carbonate sediments deposed initially : micritic laminites with ostracodes/micrite with oncoïds, in banks from 0.50 to 1 m. Fig. 2 shows a sequence of micritic laminites partially replaced by gypsum and interrupted (middle and top of the figure) by two gypsum layers in large lenticular crystals, which are themselves substituted in lutecite. La Roque-sur-Pernes quarry.

Fig. 3 Section of a micrite-oncoids sequence of which the micritic cement is almost totally replaced by microcrystalline gypsum with alabaster facies. Same location ; each subdivision on the scale represents 1 mm.

Fig. 4 In this section, an aspect of the oncoide micrites in one of the rare layers where the replacement is non-existant. These oncoids have only a weak cortication ; the nucleus is mainly a pellet. Same location.
Normal light, × 34,5.

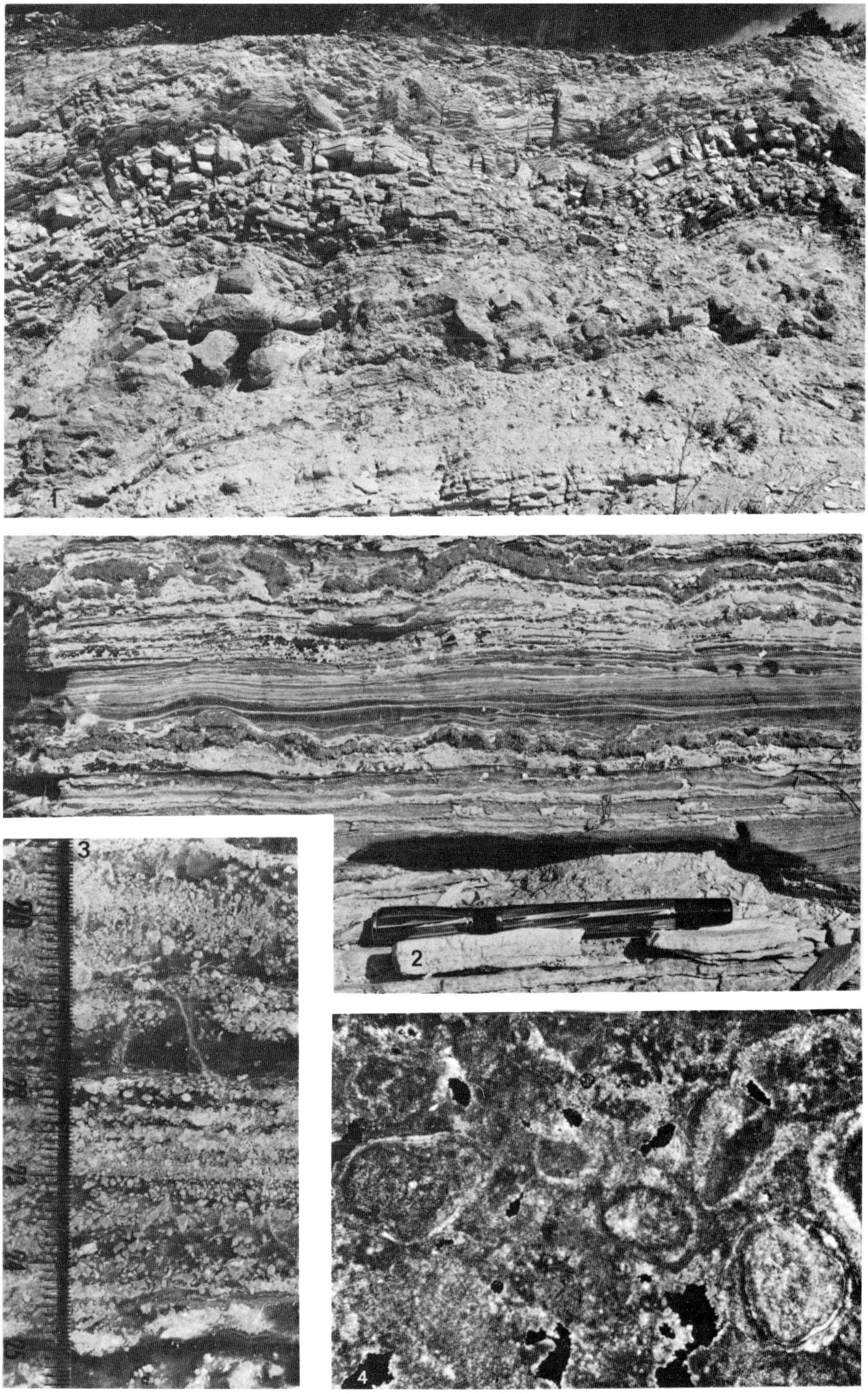
1
2
3
4

CHAPITRE XI (Planches 75 à 87). *CHAPTER XI (Plates 75 to 87)*

Evaporites d'un bassin continental subsident (Ludien et Stampien de Mormoiron-Pernes, sud-est de la France). Aspects séquentiels du dépôt; faciès primaires et leur évolution diagénétique.

Evaporites in subsident continental basin (Ludian and Stampian of Mormoiron-Pernes, in the South-East of France). Sequential aspects of deposition; primary facies and their diagenetic evolution.

PLANCHE 87 FACIES DIAGENETIQUES : SUBSTITUTIONS DE CARBONATES PAR LE GYPSE (suite et fin)

Fig. 1 et 2 Substitution partielle du ciment micritique (*cf.* pl. 86, fig. 4). Les pores créés lors de la dissolution du ciment sont remplis de gypse en monocristaux lenticulaires coalescents. Même localisation que pl. 86, fig. 2.
Lumière polarisée, × 34,5.

Fig. 3 Remplacement total du ciment micritique selon les mêmes modalités (*cf.* fig. 1 et 2). Noter l'orientation des cristaux autour de l'oncoïde et un début d'attaque du cortex de ce dernier ; au centre et à droite, mégaspore ou algue unicellulaire en sparite. Même localisation.
Lumière polarisée, × 34,5.

Dans ces trois cas, il existe deux stades au cours du remplacement : une dissolution progressive du ciment et une cristallogenèse du gypse dans les vides qui en résultent.

Fig. 4 Stade ultime du remplacement : quelques oncoïdes subsistent dans un gypse microcristallin (faciès albâtre). Le remplacement paraît plus direct et homogène. Même localisation.
Lumière polarisée, × 13,5.

Fig. 5 Les oncoïdes résistent beaucoup plus longtemps que le ciment micritique à ces phénomènes de substitution ; toutefois, dans les niveaux où la gypsification du matériau est la plus poussée, ils présentent de nombreux golfes de corrosion (flèche). Même localisation.
(Nota : sur cette lame, le gypse a été transformé en bassanite par chauffage modéré, afin d'obtenir une meilleure figuration du phénomène).
Lumière polarisée, × 46.

PLATE 87 DIAGENETIC FACIES : CARBONATE SUBSTITUTION BY GYPSUM (continued to end)

Figs. 1 and 2 Partial substitution of the micritic cement (*cf.* plate 86 Fig. 4).
The pores created at the time of the dissolution of the cement are filled with gypsum in coalescent lenticular monocrystals. Same location on plate 86, fig. 2.
Polarized light, × 34,5.

Fig. 3 Total replacement of the micritic cement in the same manner (*cf.* Fig. 1-2) ; note the orientation of the crystals around the oncoid and the beginning of an attack on the cortex of the latter. Centre and right, megaspore or unicellar algae in sparite. Same location.
Polarized light, × 34,5.

In these three cases, there are two stages during filling : a progressive dissolution of the cement, and a crystallogenesis of gypsum in the resulting empty spaces.

Fig. 4 Ultimate stage of replacement : some oncoids remain in a microcrystalline gypsum (alabaster facies). The replacement seems more direct and homogenous. Same location.
Polarized light, × 13,5.

Fig. 5 The oncoids resist much longer against these substitution phenomena than does the micritic cement ; however, in the layers where the gypsification of the material is the most advanced, they present numerous gulfs of corrosion (arrow). Same location. (Note : on this thin section the gypsum has been transformed into bassanite by moderate heating, in order to obtain a better figuration of the phenomenon).
Polarized light, × 46.

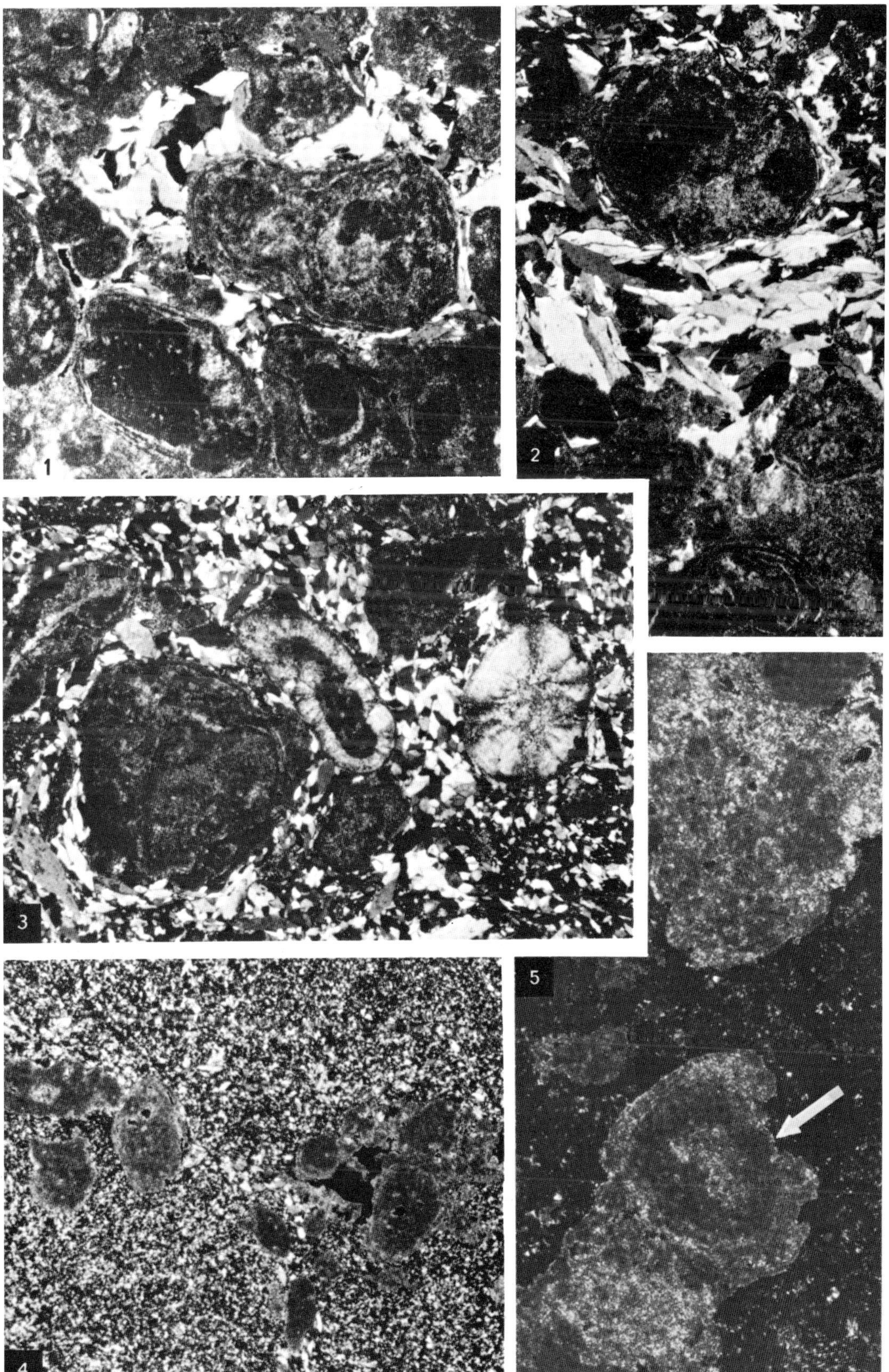
1
2
3
4
5

ALPHABETIC INDEX

ACHEVE D'IMPRIMER
EN JUILLET 1980
PAR L'IMPRIMERIE LOUIS-JEAN
05002 – GAP
N° d'éditeur 519 – N° d'impression 364
Dépôt légal : 3e trimestre 1980

IMPRIME EN FRANCE